Modern Railway Engineering Consultation

Methods and Practices

Modern Railway Engineering Consultation

Methods and Practices

Ying ZHU
Lie CHEN

China Railway Eryuan Engineering Group Co. Ltd, China

World Scientific

NEW JERSEY • LONDON • SINGAPORE • BEIJING • SHANGHAI • HONG KONG • TAIPEI • CHENNAI • TOKYO

Published by

World Scientific Publishing Co. Pte. Ltd.
5 Toh Tuck Link, Singapore 596224
USA office: 27 Warren Street, Suite 401-402, Hackensack, NJ 07601
UK office: 57 Shelton Street, Covent Garden, London WC2H 9HE

Library of Congress Cataloging-in-Publication Data
Names: Zhu, Ying, 1963– author. | Chen, Lie, 1962– author.
Title: Modern railway engineering consultation : methods and practices /
Zhu Ying & Chen Lie, China Railway Eryuan Engineering Group Co., Ltd.
Other titles: Xian dai tie lu gong cheng zi xun. English
Description: New Jersey : World Scientific, 2018 | Includes bibliographical references.
Identifiers: LCCN 2018006267 | ISBN 9789813238879 (hardcover)
Subjects: LCSH: Railroad engineering.
Classification: LCC TF146 .Y5613 2018 | DDC 625.1--dc23
LC record available at https://lccn.loc.gov/2018006267

British Library Cataloguing-in-Publication Data
A catalogue record for this book is available from the British Library.

《现代铁路工程咨询》
Originally published by China Communications Press Co., Ltd.

For any available supplementary material, please visit
https://www.worldscientific.com/worldscibooks/10.1142/10956#t=suppl

Typeset by Stallion Press
Email: enquiries@stallionpress.com

Printed in Singapore

Introduction

The authors of this book have been long engaged in the design, consultation and research of railways and directed the engineering consultation on many high-speed railways in China, including Beijing-Shanghai High-speed Railway and Wuhan–Guangzhou Passenger Dedicated Railway. This book was compiled on the basis of the systematic analysis and summary of China's railway engineering consultation and in consideration of engineering consultation practices. This book is divided into five chapters, namely 1. Development and Technical Characteristics of Railways, 2. Management of Railway Engineering Consultation, 3. Methods for Railway Engineering Consultation, 4. Key Points of Railway Engineering Consultation, and 5. Cases of Railway Engineering Consultation.

This book features excellent text descriptions and illustrations, rich contents, as well as good pertinence and practicability, so it may be used as a reference for relevant personnel engaged in the management, design, consultation and construction of railways and the teachers and students in universities and colleges.

About the Authors

Zhu Ying, born in Shanghai, is a postgraduate of Southwest Jiaotong University, Master of Engineering, the General Manager of China Railway Eryuan Engineering Group Co., Ltd., Professor-level Senior Engineer, Engineering Survey and Design Master of Sichuan Province, and an excellent expert with outstanding contributions in Sichuan Province. He has been long engaged in the survey, design and research of railway, highway and urban rail transit projects and has made remarkable achievements in such aspects as the ballastless track for high-speed railways and corresponding measuring technique, the comprehensive route selection and overall design in complex and dangerous mountainous areas, and the key technologies for major engineering construction. He directed the research on key technologies for the ballastless track test section of the Suining–Chongqing Railway (the first ballastless track test section in China), the research and innovation on the ballastless track technology for the Chengdu–Dujiangyan Railway, and the research and innovation on the precision measurement technology for ballastless track laying and maintenance of high-speed railways. He also directed the route selection and overall design for the long and main lines of high-speed railways and many trunk railways in complex mountainous areas, including the Shaoguan–Huadu section of the Wuhan–Guangzhou Passenger Dedicated Railway, the second line for the Baoji–Chengdu Railway, the second line for the Ankang–Chongqing section of the

Xiangyang–Chongqing Railway, the Chongqing–Lichuan Railway, and the Guiyang–Guangzhou Railway. He once acted as the Project Manager for the design and engineering consultation of the Wuhan–Guangzhou Passenger Dedicated Railway, the Guangzhou–Shenzhen–Hong Kong Passenger Dedicated Railway and the Shijiazhuang–Taiyuan Passenger Dedicated Railway.

Chen Lie, born in Leshan City, Sichuan Province, is a postgraduate of Southwest Jiaotong University. He is a Doctor of Engineering, the Deputy Chief Engineer of China Railway Eryuan Engineering Group Co., Ltd., and Professor-level Senior Engineer. He has been long engaged in the design and research of bridge works. In recent years, he has devoted himself to the design and consultation of high-speed railways, the design and research of long-span and steel-tied arch bridges, long-span, steel-trussed and cable-stayed bridges, and long-span concrete arch bridges for high-speed railways, research on vibration reduction and insulation technology and its standard for railway bridges, and the design and research of medium-low speed magnetic suspension bridges. He once acted as the Chief Engineer responsible for the design of the Shaoguan–Huadu section of the Wuhan–Guangzhou Passenger Dedicated Railway, the Chief Consultant responsible for the design and engineering consultation of Beijing–Shanghai High-speed Railway and many passenger dedicated railways, such as the Beijing–Shijiazhuang, Zhengzhou–Wuhan, Guangzhou–Shenzhen–Hong Kong, Shijiazhuang–Taiyuan, Hefei–Fuzhou, Shanghai–Kunming (Hunan section) and Lanzhou–Xinjiang (Gansu–Qinghai section) Passenger Dedicated Railways, as well as the Deputy Chief Consultant responsible for the engineering consultation of the Wuhan–Guangzhou Passenger Dedicated Railway.

Preface

Railways have the technical and economic advantages of convenience, time-saving, large transport capacity, safety and reliability, less land occupation, energy-saving, environmental protection, and promotion of regional economic development. Since the adoption of the *Medium and Long-term Railway Network Plan* in 2004 by the state in principle, China's railways (especially high-speed railways) have been developing quickly. The Beijing–Tianjin Intercity Rail Transit Project, the Wuhan–Guangzhou Passenger Dedicated Railway and the Zhengzhou–Xi'an Passenger Dedicated Railway, included in the first batch of railway construction projects, were put into operation in the years of 2008, 2009 and 2010 respectively. The Beijing–Shanghai High-speed Railway was put into operation in the year 2011, and many other passenger dedicated railways have also been completed and open to traffic successively. The completion of China's high-speed railway network and the realization of seamless passenger transfer among railways and other modes of transportation will certainly provide powerful transport support and stronger driving force for the economic development of China, change people's traveling habits, improve travel conditions, accelerate China's urbanization process and make the objectives of energy-saving, environmental protection and low carbon emission for transport come true.

The Ministry of Railways has introduced third-party consultation in railway engineering construction. According to the consultation contract signed by and between the consulting agency and the employer, the consulting agency is responsible for providing consultations on technical standards, technical principles, design concepts, key technologies, design documents, engineering investment and engineering construction in the processes of railway design and engineering construction, making independent consultation comments on safety, reliability, applicability, economic efficiency, durability, systematicness and interface relation of railway engineering, and systematically optimizing the engineering design by drawing on its experience, technologies and talents in railway construction, so as to ensure that safety, quality, construction period and investment of railway construction are always under control and to achieve overall construction objectives.

Railway engineering involves many disciplines and fields. In order to guarantee the consultation quality for railway engineering and meet the requirements of engineering construction, the consulting agency should attach importance to the following aspects, in addition to establishing and improving consultation management methods, developing a detailed consultation outline and mastering scientific and systematic consultation methods:

(1) In terms of general control and organization measures, emphasizing the consultants' correct understanding and knowledge about the significance and engineering characteristics of the project under consultation;
(2) Selecting the personnel (especially the key technical experts) with experience in railway design and engineering consultation to participate in the consultation, clearly specifying the consultation and review personnel for a specific project, and setting up a technical adviser and expert group to take part in the research and decision-making on major technical issues;
(3) In terms of key technologies, making pertinent scientific research and monographic study and applying the research and study results to the engineering consultation;
(4) Drawing on the consulting agency's findings in research, test, design and construction of railways as well as its advantages in

technological trends and guidelines, to apply relevant findings to the engineering consultation in time;

(5) Making systematic collection, analysis, comparison and study on relevant domestic and overseas technical standards to correctly know about the development tendency of railway technology;

(6) Enhancing technical exchanges with first-class design institutes and consulting agencies which have mature technologies and experience to know more about railway technologies and background, and conducting the consultation work in a creative way based on the actual conditions of the project under engineering consultation;

(7) Applying advanced software for structural calculation and simulation analysis to strengthen the calculation and analysis and deepen the consultations on structural details and constructions;

(8) Laying emphasis on process consultation, check and investigate on the construction site by starting from the optimization of the plane and profile of the railway line, accurately expressing consultation comments, correctly knowing about technical standards, and enhancing the review of safety and quality of works, quantities of works, design details, systems and interfaces.

To sum up, the consulting agency should conduct the consultation work with a scientific, harmonious and sustainable concept, in the principle of being scientifically rational, technically advanced and economically practical and in accordance with the requirements of building a conservation-minded society.

Of course, the effect of consultation is also related to the dominant ideas, organization and coordination capability, working methods and efforts of the employer, and the technological level of the design institute and its strength of implementing the consultation comments, in addition to the organization measures, professional skills, communication skills, service consciousness, etc. of the consulting agency.

Modern Railway Engineering Consultation — Methods and Practices was compiled by the authors on the basis of systematically summarizing and sorting the engineering consultation on the

Beijing–Shanghai High-speed Railway, the design and engineering consultations on the Wuhan–Guangzhou, Beijing–Shijiazhuang, Shijiazhuang–Wuhan, Hefei–Fuzhou, Shanghai–Kunming (Hunan section), Lanzhou–Xinjiang (Gansu–Qinghai section), Guangzhou–Shenzhen–Hong Kong and Shijiazhuang–Taiyuan Passenger Dedicated Railways, the design consultation on the Guangzhou–Zhuhai Intercity Rail Transit Project and the Wuhan Metropolitan Area Intercity Railway, and the design consultation on the Shimen–Changsha, Hankou–Yichang and Lhasa–Shigatse Railways directed by them, and in the principle of being comprehensive, systematic and practical. This book consists of five chapters. Chapter 1 first briefly introduces the overseas high-speed railway development, and then systematically introduces the speeding-up reconstruction and medium and long-term plans for China's railways, the main types of passenger dedicated railways and the construction technologies for high-speed railways, so as to allow the readers to have a systematic understanding of the development process and technical characteristics of modern railways. Chapter 2 introduces the management of railway engineering consultation, including the railway construction process and management, the organization types and staff responsibilities for engineering consultation, and the quality plan, process control, result control, organization management, technical management, working procedures and risk management for consultation. In Chapter 3, the consultation is divided into three stages, i.e. preparation, implementation and result forming, according to the working procedures and contents of consultation; the railway engineering consultation methods are classified into the investigation method, observation method, experiment method, simulation method, question-raising method, induction method, analogy method, deduction method, systematic method and feedback verification method, according to the epistemology and methodology principles for the philosophy of science and technology and in consideration with the characteristics of and actual experience in railway engineering consultation; the systematic introduction to consultation stages, definitions, classifications, characteristics and considerations applicable to these methods and the brief descriptions in combination with relevant cases are also provided. Chapter 4 introduces the bases, principles, overall requirements and key points for design optimization, construction drawing review and special consultations, and the construction process consultation for railway engineering for reference and use by the readers in railway engineering consultation. Chapter 5 lists some railway engineering consultation cases, for reference by the readers.

We hereby acknowledge Qi Baorui, the president of China Railway Eryuan Engineering Group Co., Ltd., for his guidance on and strong support for the consultation work in the charge of the authors and the compilation of this book. We would like to thank Zhang Wenjian, Min Weijing and Xu Youding for their guidance on the consultation work and their constructive suggestions. We would also like to thank Li Yuanfu and Zuo Deyuan from Southwest Jiaotong University and Wei Yongxing, Gao Jianqiang, Wang Songzhao, Hu Xinmin, Yang Gang, Liu Lirong and Yue Zhong from China Railway Eryuan Engineering Group Co., Ltd., for their review of this book. The schemes, reports and summaries of railway engineering consultations directed by us have been referenced in this book, and some figures and pictures came from relevant data and reports of the projects involved. We acknowledge the colleagues who have participated in the compilation of these data, reports and summaries.

For the purpose of correction and improvement, the readers are kindly requested to point out the improper parts in this book, since railway engineering consultation involves the newest railway technologies and many aspects such as design, construction and operation, which features strong policies, wide technical applications and numerous disciplines, and the knowledge of the authors is limited.

Ying ZHU and Lie CHEN
Chengdu, December 2012

Contents

Introduction v

About the Authors vii

Preface ix

Chapter 1. Development and Technical Characteristics of Railways **1**

1.1 Overseas High-speed Railway Development 3
1.2 Speeding-up Reconstruction of Railways in China 17
1.3 Medium and Long-term Plans for China's Railways 25
1.4 Main Types of China's Passenger Dedicated Railways 27
1.5 China's Construction Technologies for High-speed Railways 28
1.6 History of High-speed Railway Development in China 54

Chapter 2. Management of Railway Engineering Consultation **63**

2.1 Construction Process and Management of China's Railways 63

2.2 Organization Structure for Railway Engineering Consultation and Responsibilities of Key Personnel 65
2.3 Consultation Quality Plan 70
2.4 Control of Consultation Process 72
2.5 Control of Consultation Results 74
2.6 Consultation Management Systems 75
2.7 Working Procedures of Consultation 79
2.8 Consultation Risk Management 92

Chapter 3. Methods for Railway Engineering Consultation **101**

3.1 Overview 101
3.2 Stage Division of Consultation Work 102
3.3 Investigation Method 104
3.4 Observation Method 117
3.5 Experiment Method 138
3.6 Simulation Method 148
3.7 Question-raising Method 152
3.8 Induction Method 161
3.9 Analogy Method 165
3.10 Deduction Method 174
3.11 Systematic Method 176
3.12 Feedback Method 184
3.13 Miscellaneous 184

Chapter 4. Key Points of Railway Engineering Consultation **187**

4.1 Design Optimization Consultation 187
4.2 Construction Drawing Review Consultation 201
4.3 Construction Process Consultation 211
4.4 Special Consultation 216

Chapter 5. Cases of Railway Engineering Consultation 265

5.1 Consultation on Dynamic Responses of Transition Sections Among Subgrade, Bridges, Tunnels and Culverts of a Certain High-speed Railway 265
5.2 Consultation on Stress on Girder End of Intercity Railway Simply Supported Girder 308
5.3 Consultation on Improvement Design of Filling Material of Xiashu Clay for the Subgrade of the High-speed Railway 321
5.4 Consultation on Reasonable Pier Type of the High-speed Railway 326
5.5 Consultation on the River-crossing Tunnel Works of a Certain High-speed Railway 341

References **355**

Chapter 1

Development and Technical Characteristics of Railways

In 1804, Trevithick, a British engineer, built the first railway steam locomotive named "Newcastle." After its successful trial run on a circular track, this steam locomotive went from Merthyr to Abercynon, with a speed of 4 km/h. On September 27, 1825, the world's first railway that was formally put into operation was opened for traffic between Stockton and Darlington. The speed of this railway was 4.5 km/h in the initial stage and reached 24 km/h later. From the end of the 19th century to the beginning of the 20th century, railways developed quickly and greatly promoted the development of the Western capitalist economy.

In 1876, China's first railway was completed and opened to traffic between Shanghai Concession to Wusong Town, with a total length of 14.5 km. It was dismantled in the next year, due to an accident causing the death of a Chinese and consequently rousing people's great wrath not long after operations begun. On July 1, 1952, the first railway of New China, the Chengdu–Chongqing Railway, was completed and opened to traffic (see Fig. 1.1).

Since the 1960s, the new-technology railways represented by high-speed railways have emerged in the countries with developed economy and technology. Compared with other modes of transport, the transport by high-speed railways is marked by safety, reliability,

Fig. 1.1 Completion and Opening to Traffic of Chengdu–Chongqing Railway (the First Railway of New China)

technological innovation and quality service, and characterized by high speed, large passenger traffic volume, good adaptability to all weather conditions, comfort and stability, low energy consumption, light pollution, small land use and high benefit. As the development tendency of railway transport, high-speed railways have many technical and economic advantages, compared with traditional railways, including more convenience and time-savings, larger carrying capacity, better safety and reliability, less land use and more energy-saving. Thanks to such advantages, transport efficiency can be dramatically increased, the transport environment can be improved, traffic circulation can be enhanced and economic development can be promoted. Therefore, to solve the problem of large-scale population mobility, the most effective way for a country with a large population and a vast territory is to construct modern high-speed railways featuring safety, convenience, economic efficiency, environmental protection and reliability (see Fig. 1.2).

Fig. 1.2 A High-speed Railway in China

1.1 Overseas High-speed Railway Development

1.1.1 *Overview*

On December 19, 1958, the construction of the Tokaido Shinkansen Project in Japan was approved on the meeting of the Cabinet of Japan. The project was commenced on April 20, 1959. As the first high-speed railway in the world, Japan's Tokaido Shinkansen was completed and opened to traffic on October 1, 1964, launching the high-speed era of railway transport. Along with the worldwide recognition of the high-speed railway technology and its enormous effect on the economic and social development, high-speed railways developed rapidly in Asia and Europe (see Fig. 1.3).

In Asia, Sanyo Shinkansen, Tohoku Shinkansen, Joetsu Shinkansen, Hokuriku Shinkansen, Yamagata Shinkansen and Akita Shinkansen were completed and put into operation in Japan after the Tokaido Shinkansen, with a total operation length of 2,325 km. On April 1, 2004, the 330-km-long Korean Seoul–Busan High-speed Railway (KTX line) was completed and put into operation.

In Europe, the north section of TGV Sud–Est, as Europe's first high-speed railway, was competed and put into operation in France on September 22, 1981. Since then, the south section of

Fig. 1.3 A 100-series High-speed Train in Japan

TGV Sud–Est, TGV Atlantique, TGV Nord, the extension line of TGV Sud–Est, TGV Mediterranean Line and TGV Est, with a total length of 1,893 km were completed and put into operation. In June 1991, the Hannover–Wuerzburg High-speed Railway was completed and put into operation in Germany, followed by the Mannheim–Stuttgart, Hannover–Berlin, Cologne–Frankfurt and Nuremberg–Ingolstadt–Munich ICE High-speed Railways, with a total length of 1,088 km. Other high-speed railways were also completed and put into operation successively in Spain, Italy, the Netherlands, the United Kingdom and Belgium (see Fig. 1.4).

1.1.2 *Development history*

(1) Initial development stage

The period from 1964 to 1990 is the initial development stage of high-speed railways in the world. During this period, Japan completed the backbone railway network throughout the country, i.e. the Shinkansen network. Except for North America, Japan, France, Italy and Germany — which were the most economically and technologically developed countries at that time — promoted the first climax of high-speed railway construction.

The operation of Japan's Tokaido Shinkansen and France's TGV Sud–Est was a great success in the technical, commercial,

Fig. 1.4 A High-speed Railway in Germany

financial and political aspects. All investment in Japan's Tokaido Shinkansen was recovered only within 8 years, while that in France's TGV Sud–Est was recovered within 10 years.

During this period, the application of new technologies to high-speed railways not only enhanced competitiveness of railways, solved the problem of insufficient transport capacity, brought about the rise of railway passenger transport, and improved economic benefit, but also resulted in the promotion of balanced economic development in the regions along railways, the construction of relevant industries, the technological transformation of existing railway networks driven by new projects, the improvement of national existing facilities and the benefit from such improvement (see Fig. 1.5).

In the initial stage of high-speed railway construction, Japan, France, Germany and Italy spent tremendous expenses on

Fig. 1.5 A High-speed Railway Station in France

research and development and completed their own high-speed railways by using self-developed technologies.

(2) Stage of railway network planning

Japan, France and Germany set about to formulate their national master plans for high-speed railways at the initial stage of high-speed railway construction to guide the construction.

In May 1967, the Japan Society of Civil Engineers (JSCE) raised a railway network proposal, in which the north–south high-speed railway network was considered as the skeleton, with a planned total extension length of high-speed railways of 3,300 km, as shown in Fig. 1.6(a). In August of the same year, Japanese National Railways (JNR) conceived a nationwide Shinkansen railway network, with a planned total extension length of high-speed railways of 4,500 km, as shown in Fig. 1.6(b). In August of the following year, Japan Railway Construction Public Corporation (JRCC) put forward the trial proposal on a nationwide Shinkansen railway network, with 4,750 km of planned total extension length of high-speed railways, as shown in Fig. 1.6(c). In May 1969, the Cabinet of Japan approved the nationwide comprehensive development plan and planned to

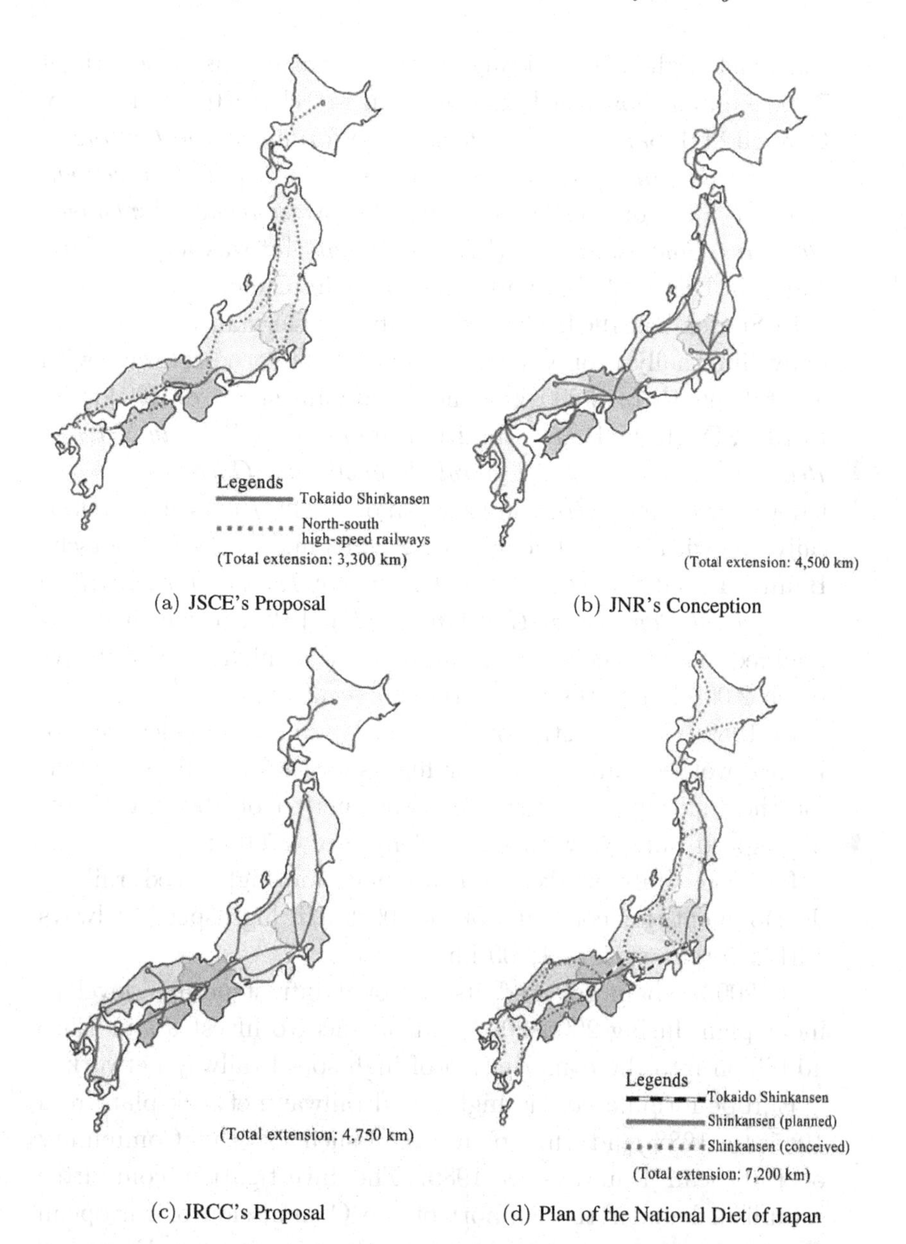

Fig. 1.6 Japan's Plans for High-speed Railways

Note: This graph is used only to show the planned routes of high-speed trains in Japan, excluding the islands.

construct high-speed railways with a total extension length of 7,200 km, as shown in Fig. 1.6(d). In March 1970, the Railway Council of Japan made the decision to formulate the *Outline of Nationwide Shinkansen Railways Construction and Improvement Act*. In May of the same year, the *Nationwide Shinkansen Railways Construction and Improvement Act* was approved by National Diet of Japan and enacted for implementation.

In September 1964, Deutsche Bahn (a German railway company) internally proposed to construct high-speed railways with a total length of 3,200 km and maximum speed of 200 km/h. In 1970, Deutsche Bahn submitted the *Plan for Railway Network Reconstruction in the Federal Republic of Germany* to the federal government, for construction of 300 km/h new high-speed railways with a total length of 2,200 km. In 1985, Deutsche Bahn released the *Development Plan for Transport Network in the Federal Republic of Germany — BVWP85*, in which it was planned to make the total operation length of high-speed railways reach 2,000 km at the end of the 20th century.

In 1989, the Ministry of Transport and Communications of France worked out its plan for high-speed railway development for the following 20 years, for construction of 300 km/h new high-speed railways with a total length of 4,700 km.

In 1986, Italy worked out its plan for high-speed railway development, for construction of 300 km/h high-speed railways with a total length of 1,200 km.

In 2000, Spain released its national infrastructure development plan during 2000–2007 and planned to invest about Euro 40 billion into the construction of high-speed railway network.

Europe formulated the high-speed railway network plan from 1985 to 1988 and this plan was issued by the Community of European Railways in 1989. The investigation committee established under the support of the Commission of European Community presented a report on the planning for European high-speed railway network in 1990 and this report was approved by the European Community Association in December of the same year. Thus, the *Master Plan for High-speed Railways in*

Europe (2010) was formed and it was planned to make the total length of high-speed railways in Europe reach 12,500 km by the year 2010. In 1993, the European Union conceived the *Pan-European Transport Networks*, which was formally approved in 1996.

In 2009, the United States House of Representatives committees issued the acts on transport and infrastructures and intended to employ USD 50 billion on preparing the six-year high-speed railway plan and developing high-speed railway passages. The long-term plan for high-speed railways of the United States is to build a high-speed railway network with a total length of 27,000 km, to adopt the most advanced track and control systems and elegant stations in various types, and to manufacture high-speed trains with a speed of 350 km/h.

(3) Forming of railway network

At the beginning of the 1990s, the countries with completed high-speed railways entered a period of high-speed railway network construction. The construction of high-speed railways was not only in demand by the railway sector, but also necessary for the interrelation of regions for political purposes. National and transnational high-speed railway networks emerged since the development of pollution-free high-speed railways was called on according to energy and environment requirements.

A nationwide high-speed railway network was formed in Japan, and Yamagata Mini-Shinkansen and Akita Mini-Shinkansen were opened to traffic in 1992 and 1997 respectively, creating a new way for the speeding-up reconstruction of existing railways. France and Germany conducted the reconstruction of existing railways in addition to building new high-speed railways.

The climax of high-speed railway construction was reached in Europe, covering France, Germany, Italy, Spain, Belgium, the Netherlands, Sweden and the United Kingdom. In October 1994, the Channel Tunnel connected France and the United Kingdom, being the first international connection line of high-speed railways. In 1997, the "Eurostar" departing from Paris connected France, Belgium, the Netherlands and Germany.

(4) Development orientations

The main orientations of overseas high-speed railway technology development can be summarized as follows:

(i) To improve the quality of track by applying ballastless tracks and continuously welded rail tracks, with constant stability of track and durability of geometric dimensions, and reduced maintenance cost (see Fig. 1.7);
(ii) To increase the speed of high-speed trains to 360–400 km/h and develop tilting trains and double-deck passenger cars;
(iii) To adopt the power distributed system for the high-speed trains with a speed higher than 250 km/h;
(iv) To reduce the axle weight of trains;
(v) To develop the hybrid brake system.

In addition, the safety assurance measure for high-speed railways, i.e. the train operation control system, is an integrated system incorporating wireless communication, satellite positioning and intelligent automatic control technologies, representing the development mainstream. The European Train Control System (ETCS) standards have been established in Europe.

Fig. 1.7 A New Type Ballastless Track on a Rail Transit Exhibition in Germany

1.1.3 *Technical characteristics*

(1) Shinkansen in Japan (see Table 1.1)

(1) The "substitution of bridges for railways" is widely applied, the bridge and tunnel ratio is increased and the track standards are improved. The bridge and tunnel ratio of the Tokaido Shinkansen opened in 1964 is 48%, while that of the Joetsu Shinkansen opened in 1982 is as high as 99%.

(2) More and more ballastless tracks have been constructed. Ballasted tracks were constructed along the whole line of the Tokaido Shinkansen in the early days, but it was hard to maintain the geometric dimensions of the tracks due to serious ballast pulverization, large subgrade settlement, and mud pumping. Since 1975, slab tracks were energetically popularized and ballastless tracks were applied to 91% of the Joetsu Shinkansen. At present, the tracks constructed mainly include anti-vibration slab tracks, slab tracks for soil subgrade, and framework slab tracks (see Fig. 1.8).

Fig. 1.8 Japanese Slab Ballastless Tracks in Tunnel

Table 1.1. Main Technical Standards for Shinkansen in Japan

Railway Line	Design Speed per Hour	Maximum Superelevation	Minimum Radius of Curve	Maximum Gradient	Type of Track
Tokaido Shinkansen	250 km	200 mm	2,500 m	15‰	Ballasted
Sanyo Shinkansen	260 km	155 mm	4,000 m, and individually 3,500 m to the west of Okayama	15‰	95%: ballasted; 5%: ballastless (to the east of Okayama) 31%: ballasted; 69%: ballastless (to the west of Okayama)
Tohoku Shinkansen	260 km	155 mm	4000 m	15‰	18%: ballasted; 82%: ballastless
Joetsu Shinkansen	260 km	155 mm	4000 m	15‰	9%: ballasted; 91%: ballastless
Hokuriku Shinkansen	260 km	155 mm	4000 m	30‰	15%: ballasted; 85%: ballastless

(3) The power distributed system is adopted for the trains to constantly reduce the axle weight and improve the train performance in an all-round way.
(4) The investment in infrastructures is reduced by improving the train sealing performance and adopting low design loads, small tunnel sections and short distances between centers of tracks.
(5) Test sections have been set up to solve the technological issues on the basis of the research on test sections. A 37-km long pilot section was built for Tokaido Shinkansen, a 260-km/h train operation test was conducted on the pilot section of Sanyo Shinkansen, and a 42.8-km long hill test section was built in the early construction stage of Tohoku Shinkansen and Joetsu Shinkansen respectively.
(6) With a fixed 6-h "possessive interval" arranged everyday, a crossover is provided between main lines in the throat areas of large and medium stations, no crossover is provided for the main lines in the throat areas at both ends of a small

station, no section crossover is provided, and a short distance between stations is adopted.

(7) Shinkansen is known as one of the safest high-speed railways and its high-speed trains are among those with the best stability worldwide, because of the high train traffic density, huge seating capacities, large passenger transport capacity, good service facilities, convenient transfer and good safety performance.

(2) TGV in France (see Fig. 1.9 and Table 1.2)

(1) The technological leading status has been kept in the traditional wheel-rail domain. On April 3, 2007, a TGV test train set the record as the fastest wheeled train in the world, reaching 574.8 km/h on the TGV East–Europe which was about to be open to traffic.

(2) The power centralized system and articulated cars are adopted for the electric multiple units (EMUs).

Fig. 1.9 A Characteristic High-speed Railway Bridge in France

Table 1.2. Main Technical Standards for High-speed Railways in France

Railway Line	Design Speed per Hour	Maximum Superelevation	Minimum Radius of Curve	Maximum Gradient	Type of Track
TGV Sud–Est	270 km	180 mm	4,000 m	35‰	Ballasted track
TGV Atlantique	300 km	180 mm	4,000 m	25‰	Ballasted track
TGV Nord	300 km	180 mm	6,000 m	25‰	Ballasted track
Extension Line of TGV Sud–Est	300 km	180 mm	North: 4,000 m; South: 6,000 m	North: 35‰; South: 25‰	Ballasted track
TGV Mediterranean Line	350 km	180 mm	7,000 m	35‰	Ballasted track
TGV Est	350 km	180 mm	West: 8,333 m; East: 6,250 m	35‰	Ballasted track

(3) The high train departure density not only provides convenience for passengers, but also induces a large passenger flow.
(4) The investment in infrastructures is reduced by improving the performance of mobile equipment.
(5) TGV lines are long. TGV Sud–Est was the only high-speed railway with a length greater than 1,000 km before the entering into operation of the Wuhan–Guangzhou Passenger Dedicated Railway in China, and the average speed of TGV trains on this railway line is over 300 km/h and the trains run stably.
(6) As the distance between stations is long, with the longest distance up to 274 km, a two-way crossover is set at every 20–25 km and large-size turnouts are used in this long-distance section to facilitate train operation dispatching.
(7) Ballasted tracks are mainly constructed.
(8) The bridge and tunnel ratio is low and the bridge styles are characteristic.

(3) ICE in Germany (see Fig. 1.10 and Table 1.3)

(1) The mixed passenger and freight transport mode is adopted. The most urgently needed high-speed railway sections are built on trunk lines first, and then these sections are connected with other sections of existing railways, forming the mode of mixing new and old railways. ICE high-speed railway network consists of the reconstructed parts of old railways (with the maximum speed of 200 km/h) and the new high-speed railways (with the maximum speed of 250–300 km/h). Most high-speed railways have to serve not only the ICE (Inter City Express) trains, but also IC (InterCity) trains, EC (EuroCity) trains, D (Durchgangszug) trains, as well as freight trains. The transport mode of passenger dedicated railway lines is partly adopted.

(2) Many new technologies are applied to the trains, including the three-phase AC drive technology, the computer-based locomotive traction control and train braking technology, the light-duty car body structure, the aerodynamics technology with low energy consumption and low noise, and the automatic diagnosis system for trains.

(3) Due to the application of mixed passenger and freight transport mode, turnouts with different sizes have been selected for high-speed railways in Germany based on the train speed per hour.

(4) Ballastless tracks are laid on the passenger dedicated railway lines. Germany started its research on ballastless tracks since the 1970s and more than 10 types of ballastless tracks have been tested. At present, six types of ballastless tracks, i.e. Bögl, Rheda, Züblin, Atd, Getmc and Berlin tracks, have been approved formally.

(5) The bridge and tunnel ratio for high-speed railways in Germany is between those in Japan and France.

Fig. 1.10 A High-speed Train in Germany

Table 1.3. Main Technical Standards for High-speed Railways in Germany

Railway Line	Design Speed per Hour	Maximum Superelevation	Minimum Radius of Curve	Maximum Gradient	Type of Track
Hannover–Wuerzburg	250 km (for passenger trains)/80 km (for freight trains)	85 mm	7,000 m	12.5‰	Ballasted track
Mannheim–Stuttgart	250 km (for passenger trains)/80 km (for freight trains)	85 mm	7,000 m	12.5‰	Ballasted track
Berlin–Hannover	250 km (for passenger trains)/80 km (for freight trains)	85 mm	4,400 m	12.5‰	Ballastless track
Cologne–Frankfurt	300 km	170 mm	3,500 m	40‰	Ballastless track
Nuremberg–Ingolstadt	300 km	160 mm	4,085 m	20‰	Ballastless track

1.2 Speeding-up Reconstruction of Railways in China

1.2.1 *Speeding-up tests and speeding-up reconstruction (see Figs. 1.11 and 1.12)*

China conducted large-scale speeding-up reconstruction of railways six times from April 1997 to 2007.

(1) Speeding-up tests

China conducted train speeding-up tests before the first large-scale speeding-up reconstruction.

In September 1995, a freight train speeding-up test was conducted on the existing Shanghai–Nanjing Railway, in which the train reached a maximum speed of 94 km/h under the traction by DF_8 and DF_{4E} locomotives. In October 1995, a passenger train speeding-up test was conducted, in which the

Fig. 1.11 A Train Running on the Railway Line after Speeding-up Reconstruction

Fig. 1.12 Speeding-up Reconstruction of Railways

train reached a maximum speed of 173.3 km/h under the traction by DF_{11} locomotives.

In November 1995, a passenger train speeding-up test was conducted on the existing Qinhuangdao–Shenyang Railway, in which the train reached a maximum speed of 175.4 km/h under the traction by DF_{11} locomotives.

From June to July 1996, a passenger train speeding-up test was conducted on the existing Shenyang–Shanhaiguan Railway, in which the train reached a maximum speed of 183.5 km/h under the traction by DF_{11} locomotives.

In November 1996, the first passenger train speeding-up test on China's electrified railways was conducted on the existing Zhengzhou–Wuhan Railway, in which the train reached a maximum speed of 185 km/h under the traction by SS_8 locomotives.

In December 1996, a passenger train high-speed test was conducted on the circular railway of Beijing, in which the train reached a maximum speed of 212.6 km/h under the traction by SS_8 locomotives.

In June 1998, a passenger train high-speed test was conducted on the existing Zhengzhou–Wuhan Railway, in which the train reached a maximum speed of 239.7 km/h under the traction by SS_8 locomotives.

(2) The first large-scale train speeding-up on existing railways
In April 1997, fast passenger trains with a speed over 120 km/h were operated on Shanghai–Nanjing, Zhengzhou–Wuhan, Shenyang–Shanhaiguan and Beijing–Qinhuangdao Railways, with the maximum train running speed reaching 140 km/h and the whole journey speed over 90 km/h, achieving the objective of departing at night and arriving in the next morning. In the whole railway network, the average speed of passenger trains reached 54.9 km/h and that of freight trains reached 31.4 km/h. Compared with the train operation diagram of 1993, the average speed was increased by 6.8 km/h and 1.4 km/h for passenger trains and freight trains respectively in the whole railway network.

(3) The second large-scale train speeding-up on existing railways
In October 1998, the objective of departing at night and arriving in the next morning was achieved in the scope of 1,500 km centered on Beijing and by focusing on the three trunk lines of Beijing–Shanghai, Beijing–Guangzhou and Beijing–Harbin Railways. The length of sections with a speed of fast passenger trains over 120 km/h was greater than 5,200 km, and the maximum speed reached 160 km/h. The "new speed" tilting trains were adopted on the Guangzhou–Shenzhen Railway, with a maximum speed of 200 km/h. The average speed of passenger trains reached 55.2 km/h in the whole railway network.

(4) The third large-scale train speeding-up on existing railways
In October 2000, the speeding-up reconstruction was carried out within a length scope of 7,200 km, mainly involving Lanzhou–Lianyungang, Lanzhou–Xinjiang, Beijing–Kowloon and Zhejiang–Jiangxi Railways. The average speed of passenger trains and freight trains reached 60.3 km/h and 32.6 km/h, respectively in the whole railway network.

(5) The fourth large-scale train speeding-up on existing railways
On October 21, 2001, the extension length of railways with train speeding-up reached 13,000 km, covering most provinces

(prefectures) and cities throughout the country. The key sections under speeding-up include those along the Beijing–Kowloon, Wuchang–Chengdu (Wuhan–Danjiang, Xiangyang–Chongqing and Dazhou–Chengdu) Railways, the southern sections of the Beijing–Guangzhou Railway, the Zhejiang–Jiangxi Railway and the Harbin–Dalian Railway. The average speed of passenger trains reached 61.6 km/h in the whole railway network.

This speeding-up further set up the railway image of passenger and freight transport: (i) the train operation schedules for departing at night and arriving in the next morning were optimized; (ii) the purpose of speeding-up is to better develop the railway traveling market; (iii) the quantity of parcel express special trains was increased and the train operation scheme was optimized; (iv) a large amount of through trains departing at regular intervals were arranged on the basis of the full investigation, analysis and sorting of bulk through freight sources and freight flows.

(6) The fifth large-scale train speeding-up on existing railways
On April 18, 2004, 19 pairs of Z-series non-stop trains were operated on the main railway trunk lines, such as the Beijing–Shanghai and Beijing–Harbin Railways. The speed of trains almost reached 200 km/h in some sections of the trunk lines, the total length of which was greater than 16,500 km. The average speed of passenger trains reached 65.7 km/h in the whole railway network.

(7) The sixth large-scale train speeding-up on existing railways
On April 18, 2007, in addition to the speeding-up of most of the existing trains, EMU trains with their train codes starting with D were added on the speeding-up railway trunk lines. The speed of passenger trains was 200–250 km/h, reaching the highest level in the international speeding-up reconstruction of existing railways. The railway trunk lines under this speeding-up reconstruction included the Beijing–Harbin, Beijing–Shanghai, Beijing–Guangzhou, Beijing–Kowloon, Lanzhou–Lianyungang, Zhejiang–Jiangxi, Lanzhou–Xinjiang, Guangzhou–Shenzhen, Qingdao–Jinan Railways. The average

speed of passenger trains reached 70.18 km/h throughout the country.

In terms of freight transport, freight trains with a speed of 120 km/h and a load of 5,000t were operated on the existing railway trunk lines under speeding-up reconstruction.

(8) The 200-km/h level speeding-up reconstruction of and comprehensive tests on Qingdao–Jinan Railway

In the 200-km/h level speeding-up reconstruction of Qingdao–Jinan Railway, the minimum radius of curve was 2,800 m for the newly-built railway sections and 2,200 m for the reconstructed sections of the existing railway lines. The existing subgrade was reinforced by driving six pieces of 30 cm^2 compaction piles into each sleeper of a single track. For bridges, wet-jointed concrete T-girders were adopted and the dynamic simulation analysis was conducted. Two fifths of existing stations were closed, with the maximum distance between stations being 24.9 km. 60 kg/mPD_3 rails, trans-section continuously welded rail tracks, type III sleepers, type I elastic rail fastenings and Nos. 12 and 18 turnouts were used. CTC centralized dispatching, GSM-R wireless communication and all auto-tensioned simple catenary suspension for overhead contact system (OCS) were adopted.

From June to July 2006, many tests were conducted, including CTC-2 train control system test on existing railways, comprehensive performance test on CHR_2EMUs, harmonic test on traction substations, comprehensive test on train crossing, function verification of centralized traffic control (CTC) system, GSM-R data communication and transmission characteristic test, test on the interference of train tracing operation on track circuit, and through turnout main test on the turnout No. 18 for passenger dedicated railways.

According to the test verification, Qingdao–Jinan Railway was capable of running 200–250 km/h EMUs, 100 km/h freight trains and double-stacked container trains with the axle load of 25t, and the complete set of technologies for 200-km/h level speeding-up reconstruction of existing railways were fully mastered.

1.2.2 *Speeding-up reconstruction technologies*

(1) Capacity-expanding reconstruction

(1) Extend the effective length of receiving-departure track. Extension of the effective length of receiving-departure track is an important measure to increase the traction mass of trains, which may greatly improve the freight transport capacity of railways.

(2) Add stations. Station addition is an effective measure to give full play to the section carrying capacity of a single-track railway.

(3) Apply double-track interpolation to the capacity control section. For an existing single-track railway, when the increased traffic demand could not be satisfied by such measures as station addition, double-track interpolation may be adopted in case of a slow traffic demand increase which requires a capacity between a single-track railway. Under the double-track interpolation option, track construction can be reduced by 50–60%, compared with the option of a completely double-track railway. However, it is more beneficial to control the service life of double-track interpolation to be more than 10 years.

(4) Perform electrification reconstruction. Electrification reconstruction of existing railways, in which the traction type is changed and high-power locomotives are used, is an effective measure to increase traction mass and running speed.

(5) Construct a second line. When the carrying capacity of a single-track railway is fully utilized and the traffic demand increases rapidly, the construction of a second line is a very effective measure to improve the carrying capacity of the railway. After a single-track railway is changed into a double-track railway completely, the carrying capacity may be increased by 1–2 times if semi-automatic block is applied. The capacity expansion will be more obvious if automatic block is applied.

(2) Speeding-up reconstruction

(1) Improve infrastructure conditions. Plane and profile conditions of the railway line may be improved by means such as adjustment of curve radius, cutoff works, track lifting and lowering. The infrastructure conditions for bridges may be improved by increasing the curve radius of bridges on curves and reducing the longitudinal gradient of track and the gradient difference between points of slope change on bridges. As a result, both the train running speed and passenger riding comfort can be improved.

(2) Improve the conditions for civil works and stations of existing railways. The improvement measures include: replacement of subgrade failing to meet the compaction density requirement; repair, reinforcement (including transverse pre-stressing reinforcement, transverse rigidity reinforcement, longitudinal reinforcement and bridge deck repair), partial renewal or complete reconstruction of old bridges and culverts not satisfying the speeding-up requirements; treatment of existing railway tunnel defects by taking the measures like grouting for leakage repair, lining replacement and stabilization of side and front slopes; capacity-expanding reconstruction of plane layout and passenger transport facilities of existing ways; addition of overpasses and underpasses (adopting underpasses in priority); addition of receiving-departure tracks for passenger trains and passenger platforms; reconstruction and expansion of passenger station buildings; application of wide sleepers to receiving-departure tracks for passenger trains at passenger stations; adjustment and optimization of the layout of freight stations; and strengthening of facilities and equipment capacity at key freight transport stations and yards.

(3) Close stations. Closing of intermediate stations with small traffic volumes, development of strategic loading and unloading points, promotion of long routing of locomotives and optimization of enterprise organization structure and labor

organization can be carried out to facilitate the optimal configuration of transport capacity resources, to improve labor productivity and promote the development of transport productivity.

(4) Develop technological innovation and comprehensively upgrade communication and signaling equipment. The new generation of China Train Control System (CTCS) system is adopted for centralized and unified dispatching of train running to ensure orderly operation of trains according to schedule. Technological innovation is achieved, namely, China has successfully developed the CTCS-2 train control system for existing railways, with proprietary intellectual property rights, realizing the target distance-speed control function. Global System for Mobile Communications-Railways (GSM-R) new technology is adopted to realize mixed passenger and freight trains running on busy trunk railway lines with speed of 200 km/h, effectively satisfying the requirements of transport management, construction maintenance and official communication and providing communication support for the facilities along the railway. Synchronous Digital Hierarchy (SDH) transmission system with large capacity and strong crossing capability is adopted, meeting the transmission requirements of service information involving signal, vehicle, electricity, electrification and informatization. V5.2 signaling interface switches and other measures are adopted for stored program control exchange, better improving the transport capacity of railways.

(5) Improve locomotive and car equipment. The complete set of technologies for design and manufacture of EMUs with a speed of 200 km/h and above and electric and diesel locomotives with high power, and the locomotives of DF_8, DF_{4E}, DF_{11} and SS_8 series feature high power and good adaptability. The independently developed HX high-power AC drive locomotive is characterized by excellent traction performance, high power, high adhesion availability, good start-up acceleration performance, high reliability, energy conservation and emission reduction.

1.3 Medium and Long-term Plans for China's Railways

Since the feasibility study on the construction of Beijing–Shanghai High-speed Railway was put on the agenda in 1990, China has spent nearly 20 years to complete the study, demonstration, testing, planning, design, construction and large-scale operation of high-speed railways. During this period, plenty of highly effective work was done and remarkable achievements were accomplished. Particularly, on January 7, 2004, the *Medium and Long-term Railway Network Plan* was discussed and adopted in principle at the executive meeting of the State Council. This plan clearly specifies the medium and long-term construction objectives and tasks of China's railway network and presents the master plan of railway network till 2020, allowing the construction of China's high-speed railways to enter the large-scale implementation stage. The *Medium and Long-term Railway Network Plan* (revised in 2008) was formally issued for implementation on October 31, 2008 with state approval.

The railway network development objectives of scale-up, structure improvement, quality upgrade, quick expansion of transport capacity and rapid increase in equipment level are specified in the *Medium and Long-term Railway Network Plan* according to the actual conditions of China's resource distribution and industrial layout and in consideration of the demands of national economic and social development. It was planned that by the year 2012, the operation length of railways throughout the country will be greater than 110,000 km, with separated passenger and freight lines on main busy trunk lines and improvement of passenger transport capacity; the passenger dedicated railway network consisting of "Four North–South and Four East–West Railways" with the target speed of passenger trains being 200 km/h and above and multiple intercity passenger transport systems will be constructed; and passenger dedicated railways with total length of 13,000 km will be completed and put into operation, including 8,000 km for trains with a speed of 300–350 km/h and 5,000 km for trains with a speed of 200–250 km/h. By the year 2020, the operation length of railways throughout the country will be greater than 120,000 km; the total length of the high-speed railway network consisting of passenger dedicated railways, intercity rail transit systems and mixed high-speed passenger and freight railways will be greater than 50,000 km, including more than 16,000 km for passenger dedicated railways; an 1–2-hour transport circle will be formed between central cities and adjacent provincial capitals and a 0.5–1-hour transport circle will formed between central cities and surrounding cities; the transport capacity will meet the demands of national economic and social development and the main technical equipment will reach or be close to the international advanced level (see Fig. 1.13).

The "Four North–South Railways" are Beijing–Shanghai Passenger Dedicated Railway, including Bengbu–Hefei and Nanjing–Hangzhou Passenger Dedicated Railways, connecting Beijing,

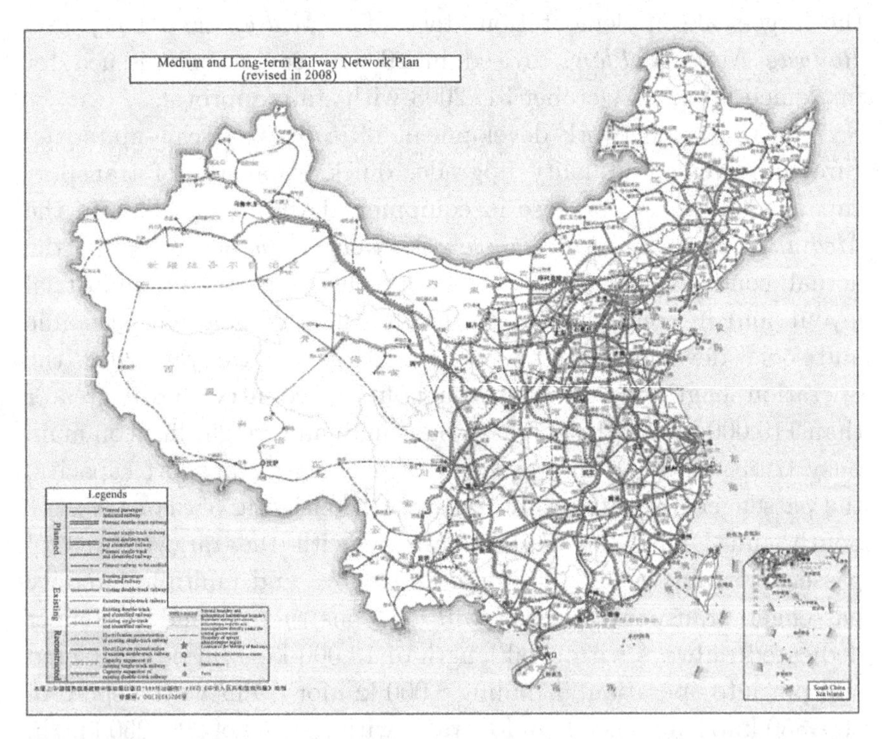

Fig. 1.13 Passenger Dedicated Railway Network Plan for China's "Four North–South and Four East–West Railways"

Tianjin and the economically developed coastal areas in the east of the Yangtze River delta; Beijing–Wuhan–Guangzhou–Shenzhen Passenger Dedicated Railway, connecting the northern and southern regions of China; Beijing–Shenyang–Harbin (Dalian) Passenger Dedicated Railway, including the Jinzhou–Yingkou Passenger Dedicated Railway, connecting the regions in the northeast and inside Shanhaiguan Pass; and the Shanghai–Hangzhou–Ningbo–Fuzhou–Shenzhen Passenger Dedicated Railway, connecting the Yangtze River and Pearl River Deltas and the southeastern coastal areas.

The "Four East–West Railways" are Xuzhou–Zhengzhou–Lanzhou Passenger Dedicated Railway, connecting the northwestern and eastern regions of China; the Hangzhou–Nanchang–Changsha–Guiyang–Kunming Passenger Dedicated Railway, connecting the southwestern, central and eastern regions of China; Qingdao–Shijiazhuang–Taiyuan Passenger Dedicated Railway, connecting the northern and eastern regions of China; and Nanjing–Wuhan–Chongqing–Chengdu Passenger Dedicated Railway, connecting the southwestern and eastern regions of China.

Multiple intercity passenger transport systems were constructed in the economically developed and densely populated regions and agglomerations, including the Circum–Bohai–Sea Region, the Yangtze River Delta Region, the Pearl River Delta Region, the Changsha–Zhuzhou–Xiangtan Region, the Chengdu–Chongqing Region, the central Henan urban agglomeration, the Wuhan urban agglomeration, the urban agglomeration in the Shaanxi plain, and the urban agglomeration on the west side of the Taiwan Strait.

1.4 Main Types of China's Passenger Dedicated Railways

(1) Long trunk railways designed for long-distance passenger transport and forming the railway network, with maximum design speed of 350 km/h, or 250 km/h in case of concurrently serving freight transport in the short term;

(2) Intercity rail transit systems designed for passenger transport between large cities or between cities in economically developed regions, with maximum design speed of 200–300 km/h;
(3) Municipal railways designed for passenger transport between the central zone of a metropolis and the surrounding population concentration areas, airports and scenic spots, with a design speed of 200 km/h in general.

1.5 China's Construction Technologies for High-speed Railways

1.5.1 *Foundational engineering technologies*

Through the mastering of analysis instruments, change rules and control measures for structural deformation, settlement and dynamic characteristics as well as emphasis on systematicness and interface management, China has built foundational engineering technologies for high-speed railways, complete engineering technologies involving route, ballastless tracks, continuously welded rail tracks, precision measurement, subgrade, bridges, culverts and tunnels of high-speed railways.

(1) Route selection technologies (see Fig. 1.14)

(i) Complete route selection technologies for high-speed railways include the route selection technologies for high-speed railways in the areas with special topographical, geological and environmental conditions, such as geologically complex and dangerous mountains, marine soft soil sediment, developed economy, densely-distributed transport corridors, collapsible loess, karst, densely-distributed river network, regional settlement, high-intensity earthquakes and cold.
(ii) The dynamic response can be reduced and the passenger riding comfort can be guaranteed by taking such measures as selecting the larger curve radius, laying emphasis on the combination between plane and profile of the railway

Fig. 1.14 A High-speed Railway

line, and increasing the distance between vertical curve and transition curve.

(iii) The maximum gradient of sectional main line is determined according to topographical conditions, checking calculation results of EMU traction power and technical and economic comparisons. Under the condition of a small amount of increase in quantities of works, a smaller gradient and a longer length of grade section are adopted in priority based on the topographic conditions.

(iv) The theories and methods of high-speed train/track coupling dynamics are used to carry out the whole-process dynamic simulation analysis, which aims to simulate and analyze the plane and profile design parameters of the railway line under high-speed running conditions and the comfort and safety indexes of subgrade and bridges, subgrade and culverts, subgrade and tunnels, bridges and tunnels, turnouts and sectional tracks, ballasted and ballastless tracks on running vehicles, evaluate the rationality of wheel-rail dynamic safety indexes (wheel-rail lateral force, derailment coefficient, wheel load reduction rate and increase of dynamic track gauge), comfort indexes (vertical

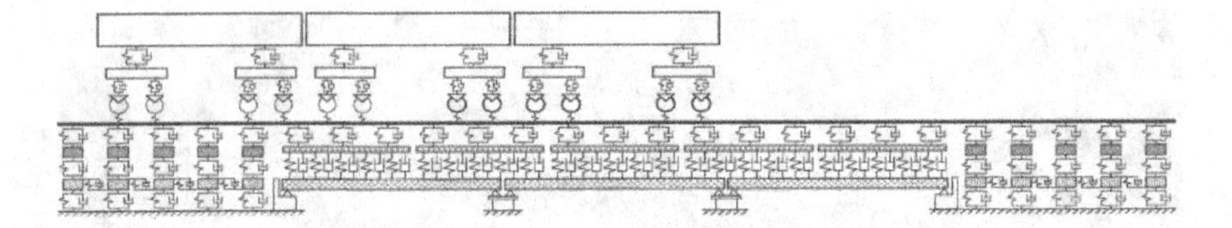

Fig. 1.15 High-speed Train/Track Coupling Dynamics

and lateral vibration accelerations and stability of the car body), plane and profile design parameters of the railway line, track structure and civil works, and optimize the plane and profile design of the railway line (see Fig. 1.15).

(2) Precise surveying network, settlement transformation calculation and observation evaluation technologies for ballastless tracks

(i) Both horizontal control network and vertical control network are divided into the survey control network, the construction control network, and operation and maintenance control network.

(ii) The horizontal control network is set at three levels based on the frame control network CP0 established by using the GPS surveying method. Level I is the foundation horizontal control network CPI, mainly providing coordinate datums for survey, control, operation and maintenance. Level II is the railway horizontal control network CPII, mainly providing control datums for survey and construction. Level III is the track control network CPIII, mainly providing coordinate datums for track laying and operation and maintenance (see Fig. 1.16).

(iii) The vertical control network is set at two levels. Level I is the benchmark control network of the railway, providing height datums for survey, design and construction. Level II is the track control network CPIII, providing height datums for track laying and operation and maintenance.

(iv) To ensure the high-regularity measurement and accurate adjustment technology of tracks, relevant survey

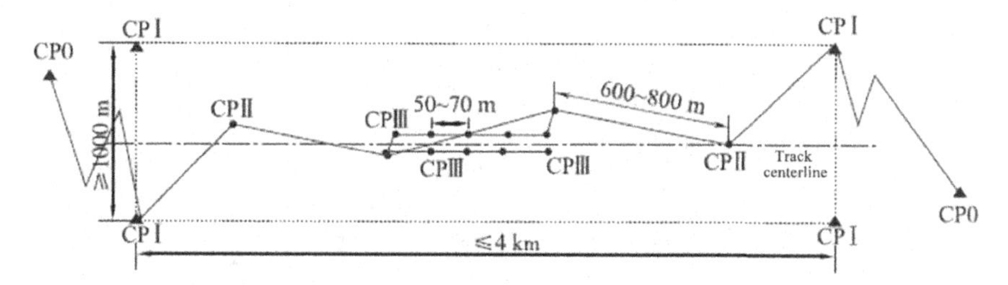

Fig. 1.16 Schematic Diagram of Three-level Horizontal Control Network for High-speed Railways

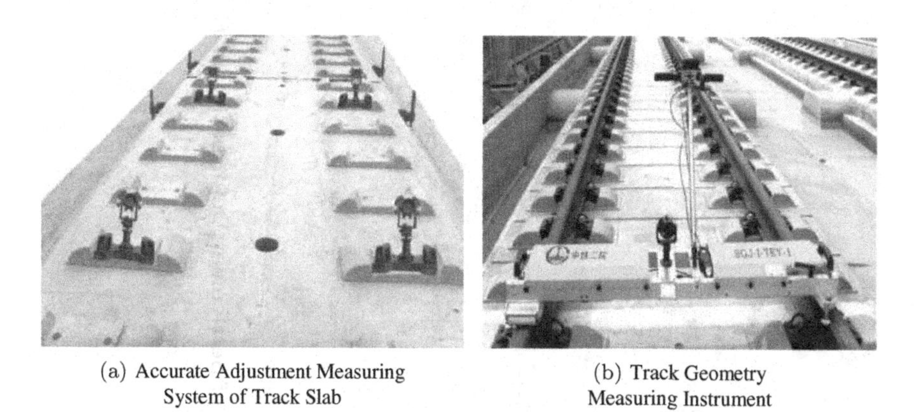

(a) Accurate Adjustment Measuring System of Track Slab

(b) Track Geometry Measuring Instrument

Fig. 1.17 Precision Measurement of Ballastless Tracks for High-speed Railways

supporting software has been developed based on the establishment of the three-level horizontal and vertical control survey network matching with high precision requirements of tracks (see Fig. 1.17).

(v) Observation and evaluation of continuous settlement transformation are carried out on civil structures including subgrade, bridges, culverts and tunnels in different construction stages to master the settlement transformation rules of structures under different geological conditions and engineering measures, determine the conditions for laying ballastless tracks and control the settlement after laying ballastless tracks (see Fig. 1.18).

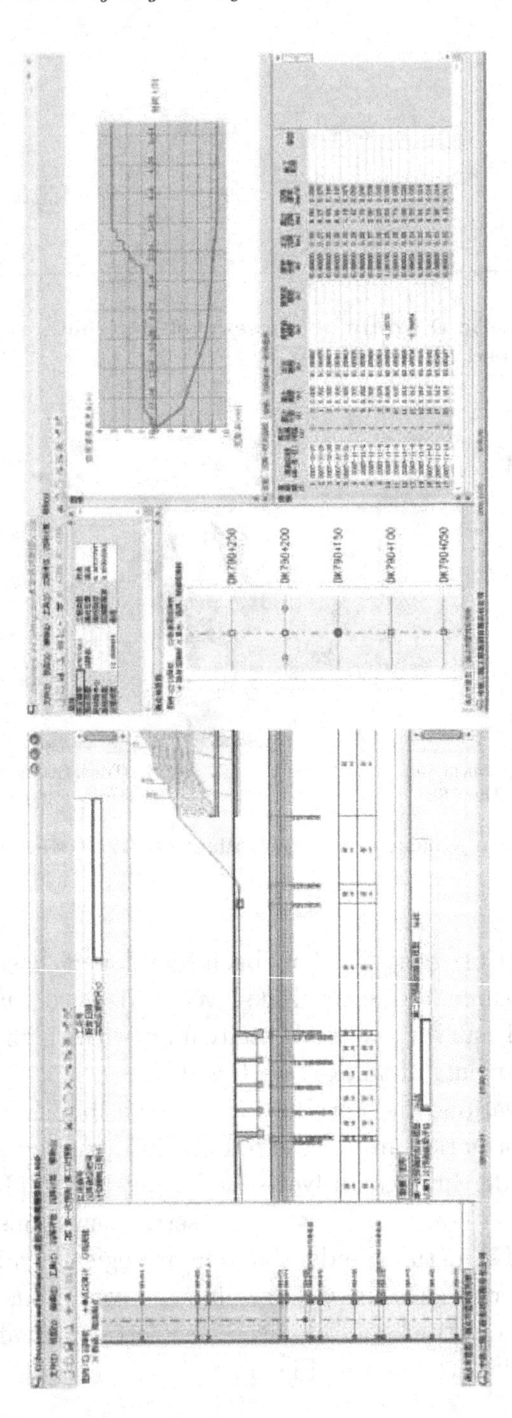

Fig. 1.18 Evaluation and Analysis of Settlement Deformation of High-speed Railway

(3) Ballastless track technologies

(i) Design and construction technologies of China Railway Track System (CRTS) ballastless tracks with proprietary intellectual property rights, including the technologies of unit slab type, longitudinally-connected slab type, double block type, and turnout sleeper buried type ballastless tracks (see Fig. 1.19);

(ii) Technologies of laying ballastless tracks on long-span concrete bridges and steel bridges (see Fig. 1.20);

(iii) Integrated grounding system technologies of ballastless tracks.

(a) Framework Slabs of Bridge Ballastless Tracks for Suining–Chongqing Railway

(b) Longitudinally-connected Slabs of Bridge Ballastless Tracks for Suining–Chongqing Railway

(c) CRTSIII Slab Ballastless Tracks on Bridge

(d) Transition between Ballasted and Ballastless Tracks

Fig. 1.19 Several Typical Track Structures

Fig. 1.20 Suining–Chongqing Railway Bridge with a Main Span of 168 m for Laying Ballastless Tracks

(4) Subgrade technologies

(i) Complete design and construction technologies of subgrade include the design and construction technologies for high-speed railway subgrade in the areas with special topographical, geological and environmental conditions, such as geologically complex and dangerous mountains, marine soft soil sediment, collapsible loess, karst, regional settlement and cold (see Fig. 1.21).

(ii) The dividing height between subgrade and bridge is determined through technical and economic comparison based on the embankment and subsoil conditions, nature and source of filling materials, local land resources and urban transportation requirements. The dividing height between subgrade and bridge may be 7–8 m for general sections, about 5 m in suburb areas, and 4–6 m for the sections with soft subsoil. Bridges should be constructed in the

(a) Subgrade

(b) Retaining Wall between Anchored Piles

(c) Slope Protection with Cutting Supporting Seeping Groove + Herringbone Water-interception Skeleton

(d) Pile-plank Wall

Fig. 1.21 High-speed Railway Subgrade

sections with soft subsoil, where there are concentrated ponds and roads, densely-distributed ditches and canals and difficulties in settlement control.

(iii) High-quality filling materials are used and the control of subgrade filling is strengthened to reduce the transformation of embankment and meet the requirements of high-speed train running and laying ballastless tracks.

(iv) With the application of composite subgrade of Cement Fly-ash Gravel (CFG) pile, pile-plank subgrade and pile-network structure, it is able to control the high-speed railway subgrade settlement for laying ballastless tracks under the conditions of deep soft embankment, subgrade in turnout area and cutting with soft subsoil (see Fig. 1.22).

(v) The structure and construction measures for bridge–road transition section, culvert–road transition section, tunnel–road transition section, embankment-cutting transition

Fig. 1.22 CFG Piles for High-speed Railway Subgrade

section and short subgrade transition section between bridges and tunnels are determined according to the longitudinal rigidity matching conditions of different civil foundations.

(vi) A subgrade settlement and transformation monitoring system is set up to provide reliable decision-making bases for settlement evaluation and information-based construction.

(vii) Subgrade and Electric and Mechanical Works (E&M) equipment are co-designed and relevant conditions are reserved for E&M works, such as track passing, grounding, OCS post, to avoid secondary excavation of subgrade.

(5) Bridge technologies

(i) Complete design and construction technologies for bridges include the design and construction technologies for high-speed railway bridges in the areas with special topographical, geological and environmental conditions, such as geologically complex and dangerous mountains, marine soft soil sediment, densely-distributed transport corridors, collapsible loess, karst, densely-distributed river network, regional settlement, high-intensity earthquakes and cold.

(ii) Strictly control the self-vibration characteristics of bridge structures, carry out train–track–bridge coupling dynamics simulation analysis as per dynamic parameters of high-speed trains, irregularity spectrum of high-speed railway tracks and 1.2 times the design speed, and make a checking calculation on the evaluation indexes, such as derailment coefficient, wheel load reduction rate, vibration acceleration of car body and riding comfort of the train, so as to ensure that all technical indexes meet the requirements for train safety and passenger riding comfort (see Fig. 1.23).

(iii) Transformation of bridge structures, especially the transformation caused by shrinkage, creepage and temperature effects after track laying, is controlled strictly.

(iv) The settlement of bridge foundation is controlled strictly. The settlement and the settlement curve of piers and abutments are calculated in accordance with the modulus and coefficient of compression under multistage pressures determined in geological test and the loading conditions for bridge pile-foundation construction of bridges. The settlement of piers and abutments is controlled by rational setting and adjustment of the number and length of piles and verification with on-site pile test.

Fig. 1.23 Through Type Steel Tied Arch Bridge for High-speed Railway

(a) Prefabrication of Box Girder

(b) Transportation of Box Girder

(c) Erection of Box Girder

(d) Pouring of Box Girder on Mobile Formwork

Fig. 1.24 Construction of Full-span Box Girder for High-speed Railway

(v) The main girder type is 32 m prestressed concrete double-track and full-span box girder and the construction is dominated by concentrated prefabrication in girder yard, transport by 900*t* girder trucks and erection at bridge site.

(vi) For sections with dense bridges and tunnels, transport of flange plate segments cut from a full span box girder through tunnels and cast-*in-situ* full span box girders by using movable formwork, false work and full hall support are applied according to topographical and engineering conditions (see Fig. 1.24).

(vii) For turnout beams of ballastless tracks, continuous structures with a proper span are selected following the principles that continuous bridge structure should be used within a certain scope of the turnout structure and its switch rail and point rail and the setting of bearing or support at the positions of switch rail and point rail should be avoided. In addition, the vertical stiffness and change in stiffness are strictly controlled for the beam body, consideration is given to the bridge temperature span and the setting of rail temperature expansion joint is avoided. The "vehicle-turnout-bridge-pier integration" calculation model and software for continuously welded turnout on bridge are used for conducting the coupling dynamic analysis of vehicles, rails, turnouts, bridges and piers and verifying the adaptability of the continuously welded turnout on bridge to the passing of high-speed trains (see Fig. 1.25).

(viii) The bridge pier body in a seismic area is provided with surface protection reinforcement to enhance the reinforcement in bearing platform of pile foundation. Seismic

Fig. 1.25 Continuously Welded Turnouts on Ballastless Tracks for High-speed Railway

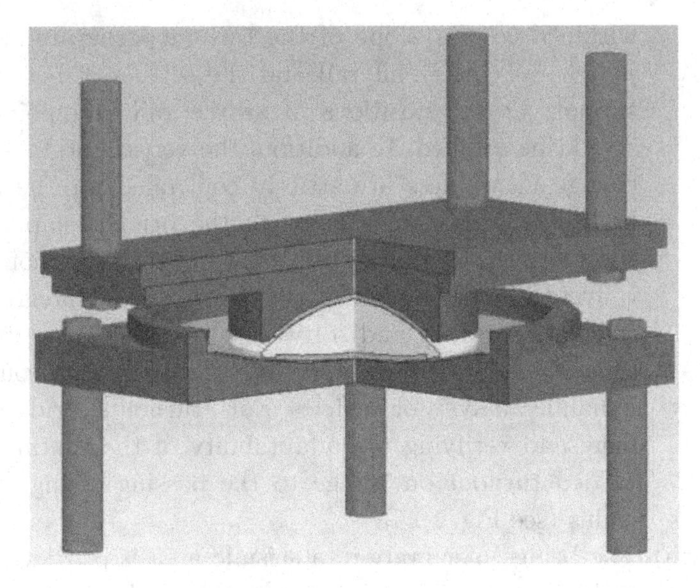

Fig. 1.26 Schematic Diagram of Friction Pendulum Bearing for Bridges in Highly Seismic Areas

mitigation and absorption technologies are adopted for bridges in highly seismic areas (see Fig. 1.26).

(ix) Frame culverts are adopted, settlement and transformation of culverts are strictly controlled and transition sections are provided between culverts and subgrade (see Fig. 1.27).

(6) Tunnel technologies

(i) Complete design and construction technologies of tunnels include design and construction technologies for high-speed railway tunnels in the areas with special topographical and geological conditions, such as geologically complex and dangerous mountains, collapsible loess, karst and rich water, high-intensity earthquakes, non-coal measure strata with high content of gases and cold (see Fig. 1.28).

(ii) Double tracks in a single tunnel are generally adopted. The cross section of tunnel is large and the maximum excavation area of a double-track tunnel usually exceeds

Fig. 1.27 Culvert for Ballastless Tracks

Fig. 1.28 High-speed Railway Tunnel

160 m^2. The cross section of tunnel is determined based on such factors as tunnel construction clearance, number of tracks and distance between centers of tracks, tunnel equipment space, reserved space, type of rolling stock and its airtightness, and the sectional area necessary for reliving aerodynamic effect. Among which, the last factor has a control effect.

(iii) Aerodynamic effects and measures like tunnel pressure change when a train passes the tunnel, micro-pressure wave, crossing pressure change, effect of longitudinal gradient on pressure change, passenger riding comfort and harm to passengers' health are considered during tunnel design (see Fig. 1.29).

(iv) The type and length of buffer structures at tunnel portals are determined based on environmental conditions at tunnel portals, tunnel aerodynamic analysis, anti-noise measures and portal design, or the effect of micro-pressure wave may be relieved by utilizing existing service gallery. The types of buffer structures at tunnel portals mainly include extended open cut tunnel and aerodynamic opening in existing open cut tunnel (see Fig. 1.30).

(v) The structural type of tunnel portal is based on the topographical and environmental conditions at the tunnel portal and consideration is also given to ecological, environmental protection and landscaping requirements in addition to safety and stress requirements (see Fig. 1.31).

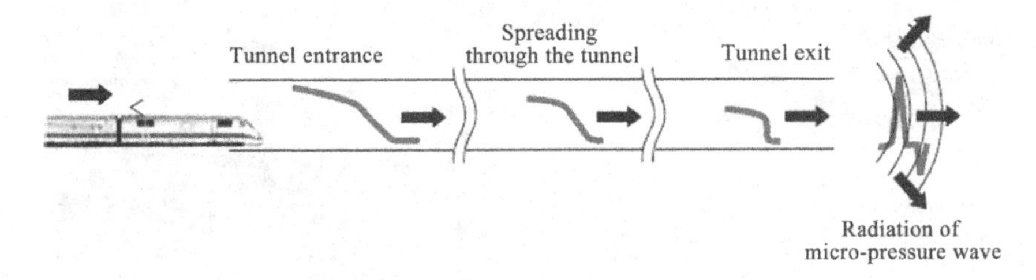

Fig. 1.29 Aerodynamic Effect in High-speed Railway Tunnel

Fig. 1.30 Effect Picture of Buffer Structure at Tunnel Portal

Fig. 1.31 Effect Pictures of Tunnel Portal Types under Different Topographical Conditions

(vi) Emphasis on waterproofing of primary support, main task of self-waterproofing of concrete structures and focus on waterproofing of construction and deformation joints are implemented according to topographical, geomorphic and hydrogeological conditions of the project and following the principles of combination of waterproofing, drainage, interception and blockage, adaptation to local conditions, and comprehensive treatment.

(vii) Composite lining is applied, composite reinforced lining is adopted at the connection between the sections, such as portal section, shallowly-buried section and fault fracture zones, and tunnels or service galleries, and open cut tunnel lining is adopted for open cut sections.

(viii) For tunnel construction, full-face tunneling method, bench tunneling method, Center Diaphragm (CD) method, Cross Diaphragm (CRD) method or double side drift method may be selected based on the tunneling cross section, geological conditions of surrounding rocks, construction period, tunnel length, etc. Smooth blasting is adopted for tunnel excavation and wet shotcreting is adopted for primary support (see Fig. 1.32).

(ix) Advanced prediction and geophysical prospecting at tunnel bottom are adopted for tunneling in the sections with soluble rocks.

(7) Station technologies (see Figs. 1.33–1.35)

(i) The architectural image of station building integrates local features, historic charm and modern characteristics in consideration of passenger flow line, functional layout, large-space structure and the requirements for saving land, energy, water and material, as well as environmental protection.

(ii) Various types adopted include aboveground entering and underground exiting of passengers, elevated waiting

Fig. 1.32 Large Cross Section Tunneling

(a) Exterior Picture

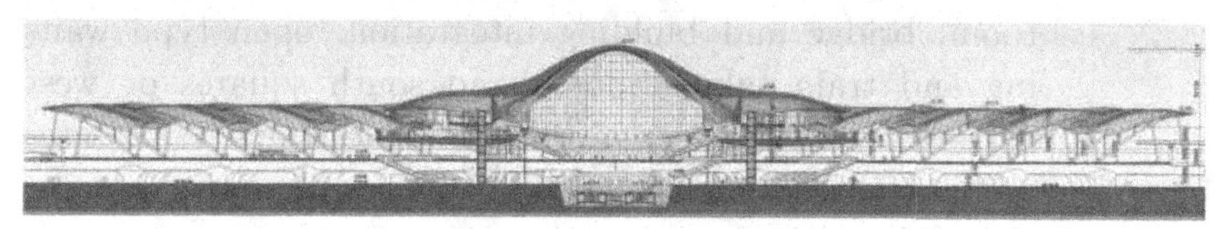

(b) Elevation Structure

Fig. 1.33 New Wuhan Railway Station

(a) Exterior Picture

(b) 3D Structure

Fig. 1.34 Chengdu East Railway Station

room, bridge and building integration, open type waiting and train taking, north and south squares or west and east squares, separation of pedestrians and automobiles, three-dimensional zero-transfer, non-post canopy, elevated station, underground station, automatic fare collection system, high and large space, energy-saving and environmental protection, automatic and barrier-free waiting room.

(iii) Long-span steel frames, A-shaped towers and canopies with long span and suspension beam and without station post

Fig. 1.35 Station Platform

are adopted to provide the passengers with capacious and bright space.

(8) Interface technologies

(i) Emphasis is laid on the design coordination between civil works and the compatibility of deformation among structures during the design of subgrade, bridges and culverts and tunnels to avoid frequent transition between different structures and ensure the evenness of track rigidity and the smooth rigidity transition between different track structures.

(ii) Emphasis is laid on the coordination between civil works and OCS, communication and signal works to realize system optimization. The design of auxiliary works for subgrade, bridges and culverts and tunnels meet the layout requirements for cable troughs, OCS, sound barriers, integrated ground wires, route signs, pipeline crossing tracks in station areas, as well as the equipment for track crossing of traction power transformation, power supply, communication and signaling cables.

(a) Welding Base of Long Rails

(b) Transport of Long Rails

Fig. 1.36 Welding and Transport of 500-m-long Rails

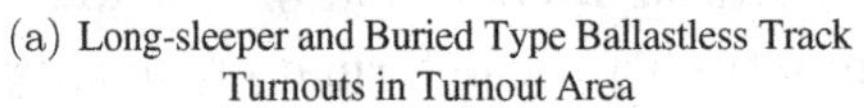

(a) Long-sleeper and Buried Type Ballastless Track Turnouts in Turnout Area

(b) Slab Type Ballastless Track Turnouts in Turnout Area

Fig. 1.37 Ballastless Track Turnouts for High-speed Railways

(9) Trans-section continuously welded rail track

(i) 100-m-long rails are adopted and welded into 500-m-long rails in plant before being transported to the site, where the rails are welded into trans-section continuously welded rail tracks (see Fig. 1.36).

(10) High-speed turnout technologies (see Fig. 1.37)

(i) The design, manufacture, transport, installation, commissioning, stress relief, maintenance and repair technologies of turnout No. 18 with a speed of 250 km/h and 350 km/h for connecting main tracks and receiving-departure tracks.

(ii) The design, manufacture, transport, installation, commissioning, stress relief, maintenance and repair technologies of turnout No. 42 with a speed of running through turnout main being 350 km/h and a speed of running through turnout branch being 160 km/h for crossovers and up and down high-speed railways.

(iii) The design, manufacture, transport, installation, commissioning, stress relief, maintenance and repair technologies of turnout No. 62 with a speed of running through turnout main being 350 km/h and a speed of running through turnout branch being 220 km/h for up and down high-speed railways.

1.5.2 *Technologies of four electric systems (traction feeding, power supply, signaling and communication systems)*

(1) Electrification technologies

(i) The single-phase power frequency (50 Hz) 25 kV AC system is adopted.

(ii) For the traction feeding system, Auto Transformer (AT) feeding system ($2 \times 25\,\text{kV}$) is applied to main tracks, while direct feeding system (25 kV) with return wire is applied to station tracks, branch tracks and connection tracks.

(iii) Auto-tensioned elastic catenary suspension and auto-tensioned simple catenary suspension are adopted for the OCS (see Fig. 1.38).

(2) Communication, signaling and train control technologies (see Fig. 1.39)

(i) GSM-R digital mobile communication system for achieving mobile voice communication and wireless data transmission.

(ii) CTC transport dispatching and commanding system for centralized dispatching and control of trains.

(iii) CTCS-2 overspeed protection system meeting the demand of operation control for high-speed railway trains with a

Fig. 1.38 OCS of High-speed Railway

speed of 250 km/h, which is also used as an equipment subsystem on high-speed railways with a design speed of 350 km/h and adopted for control of trains in deceleration operation with a speed of 300 km/h in case of wireless transmission channel failure.

(iv) CTCS-3 overspeed protection system meeting the demand of operation control for high-speed railway trains with a speed of 350 km/h, which is built in with redundant CTCS-2 functions to meet the demands on speed control of EMUs in up operation and with a speed of 250 km/h and CTCS-3 trains in down operation and with wireless transmission channel failure.

(3) Passenger service system technologies

The passenger service system for high-speed railways can perform the functions of automatic ticketing, automatic fare collection, releasing passenger service information, public address, guidance, inquiry and help, realize centralized management, system integration, information sharing, joint operation and control and emergency linkage of railway passenger service, and provide

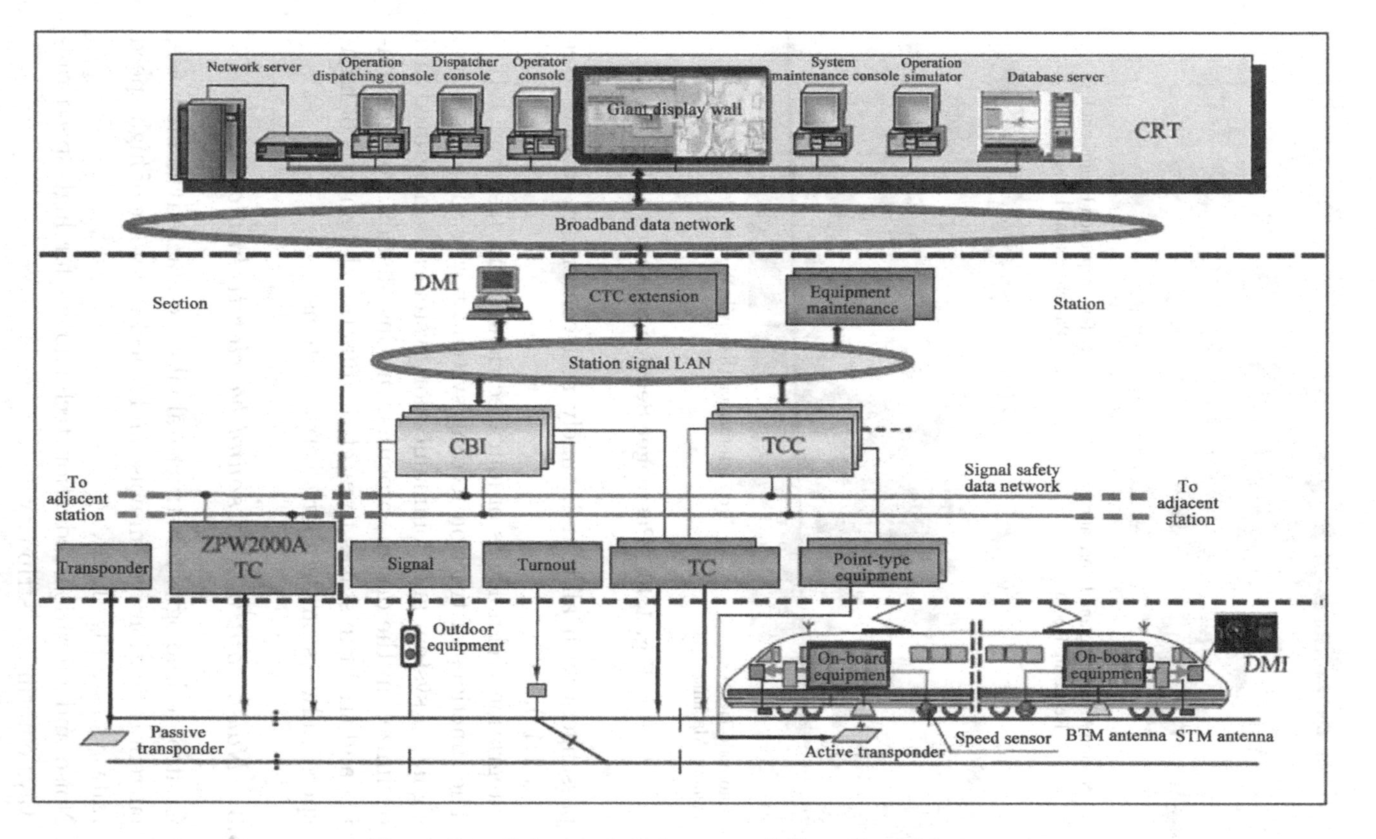

Fig. 1.39 Schematic Diagram of Signaling System

(a) Station Monitoring and Information Releasing Room

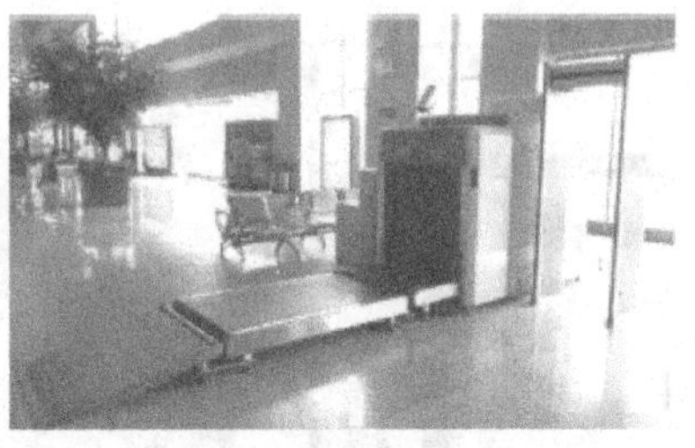

(b) Safety Inspection Equipment for Passengers and Luggage

(c) Automatic Ticket Vending Machine

(d) Information Display in Waiting Room

Fig. 1.40 Passenger Service System

passengers with all-round, timely and convenient services (see Fig. 1.40).

(4) Integration technologies of four electric systems (see Fig. 1.41) The functions of traction feeding system, power supply system, signaling system and communication system are integrated and optimized in the design, procurement, installation, commissioning and trial run stages to meet the functional demands of high-speed railways on the four electric systems.

1.5.3 *Noise control and sound barrier technologies*

(1) Control technologies for structural vibration, aerodynamic noise, current collection system noise and wheel-rail noise of high-speed railways (see Fig. 1.42).
(2) Noise and vibration reduction technologies for high-speed railways entering a "strip city."

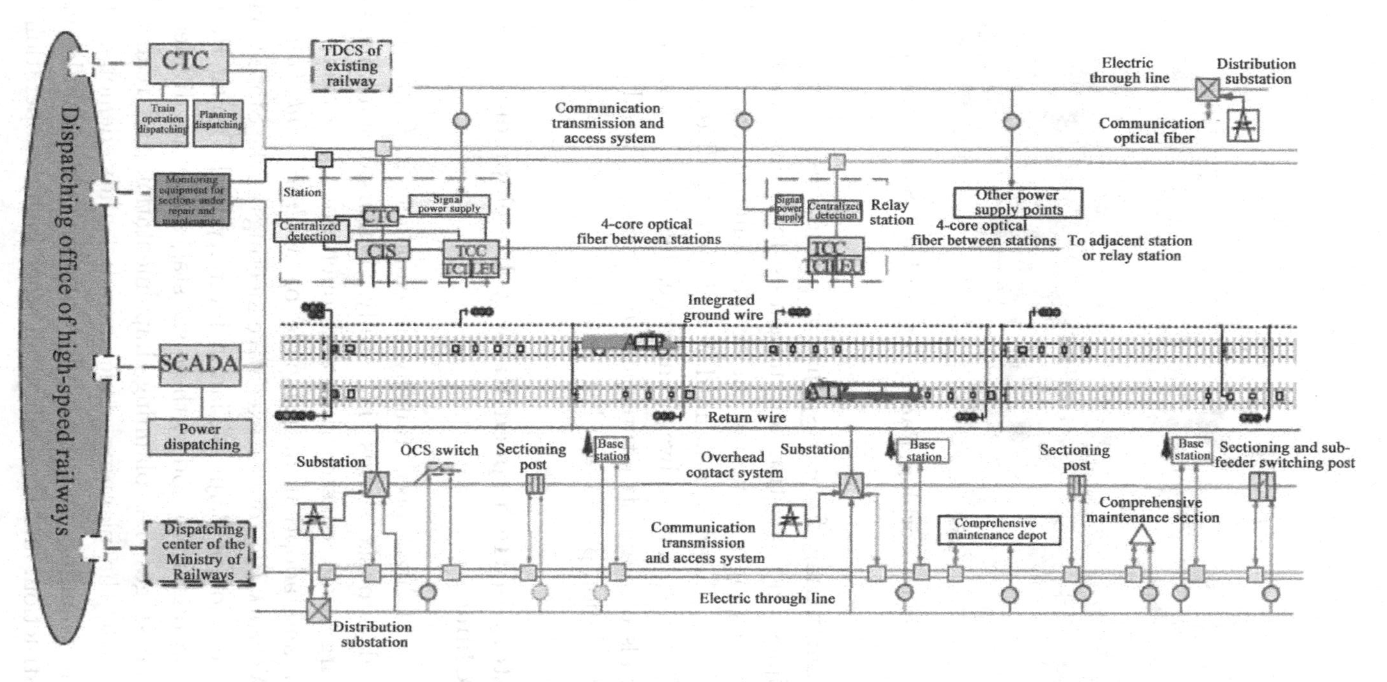

Fig. 1.41 Schematic Diagram of Integration of Four Electric Systems

Fig. 1.42 Sound-absorbing Panels for Railway Sound Barriers and Track Bed

(3) The fluctuating force response and the design parameters, principles and construction measures for sound barriers are determined by conducting an aerodynamic analysis based on train speed, shape of train head end, air density, distance between sound barrier and railway centerline, height of sound barrier, materials of main structures and materials of sound absorbing and insulation panels (see Fig. 1.43).

1.6 History of High-speed Railway Development in China

The history of high-speed railway development in China can be summarized into six stages, i.e. study, demonstration, planning, testing, large-scale construction and operation. Considering the discussion and adoption in principle of *Medium and Long-term Railway Network Plan* on the executive meetings of the State Council and the operation of Beijing–Tianjin Intercity Rail Transit Railway as two important time nodes, these six stages may be divided into three major stages, i.e. planning, implementation and realization. Among which, the planning stage includes study, demonstration, testing and planning, and the implementation stage includes large-scale construction, and the realization stage includes operation.

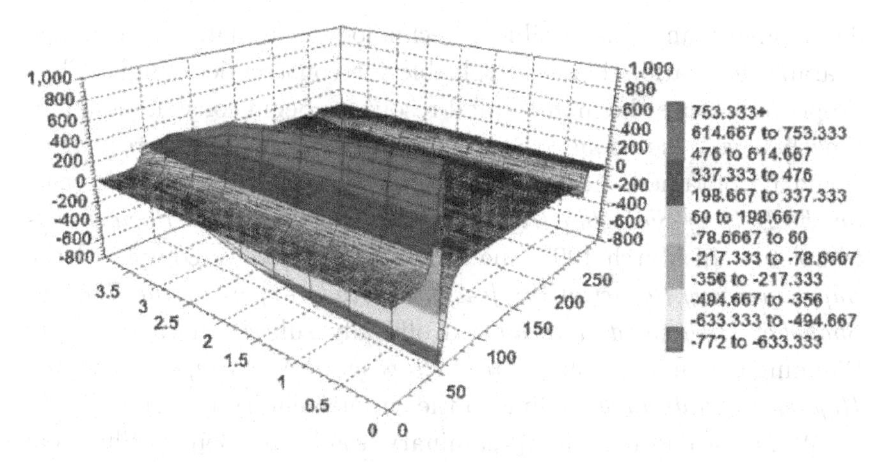

Fig. 1.43 Vibration Wave of Sound Barriers When a High-speed Train Passes

(1) Planning stage

In 1990, the relevant feasibility study about the construction of Beijing–Shanghai High-speed Railway was put forward on the State Council agenda, and in December of the same year, *Concept Report on Route Plan of Beijing–Shanghai High-speed Railway* was completed which started the history of high-speed railway development in China.

In April 1991, *Study Report on the Planning Program of Beijing–Nanjing High-speed Passenger Transport System* and *Study Report on the Planning Program of Shanghai–Nanjing High-speed Passenger Transportation System* were completed. In May 1992, after investigation and study for almost one year, China Academy of Railway Sciences submitted *Feasibility Study Report for Beijing–Shanghai High-speed Railway*. At the end of 1994, the Ministry of Railways, together with the former State Scientific and Technological Commission, State Planning Commission, State Economic and Trade Commission and State Commission for Restructuring the Economic System completed the *Preliminary Study Report on Major Technical and Economic Issues Related to the Beijing–Shanghai High-speed Railway*, in which it was considered that the construction of *Beijing–Shanghai High-speed Railway* was urgently necessary from the view of realistic development, and was feasible in technology, reasonable in economy, bearable for national power, solvable for construction funds. They proposed to seize the moment and resolve to begin construction, and they believed the sooner the construction began, the more benefit would be obtained. In April 1996, the *Pre-feasibility Study Report on the Beijing–Shanghai High-speed Railway* (*Draft for Review*) was completed. In March 1997, the Ministry of Railways completed the *Supplementary Report of Pre-feasibility Study Report on the Beijing–Shanghai High-speed Railway* and officially submitted to the State Planning Commission the *Project Proposal of New Beijing–Shanghai High-speed Railway* according to the supplementary report.

While promoting the preliminary study of Beijing–Shanghai High-speed Railway on April 13, 1999, the State Council approved the *Feasibility Study Report* on Qinhuangdao–Shenyang Passenger Dedicated Railway submitted by State Planning Commission. On August 16, 1999, Qinhuangdao–Shenyang Passenger Dedicated Railway, as the first experiment line of High-speed Railways in China, was commenced at full capacity. Qinhuangdao–Shenyang Passenger Dedicated Railway is a double-track electrified railway mainly for passenger transport with a design speed of 200 km/h. The infrastructure is designed with the conditions for speeding up to 250 km/h in the future, and the line is 407 km long. On June 16, 2002, the track-laying of the whole Qinhuangdao–Shenyang Passenger Dedicated Railway line was completed. On December 13, 2002, the test speed of "China Star" train manufactured by China on Qinhuangdao–Shenyang Passenger Dedicated Railway reached 321.5 km/h. On October 12, 2003, Qinhuangdao–Shenyang Passenger Dedicated Railway started official operations (see Fig. 1.44).

In December 1999, after one year and two months' evaluation, China International Engineering Consulting Corporation approved the *Project Proposal of New Beijing–Shanghai High-speed Railway* and considered the construction of Beijing–Shanghai High-speed Railway necessary, and its construction scheme feasible, investment scale reasonable, and economic benefit feasible. They believed it was time to seize the moment to initiate the project as soon as possible. In May 2000, China International Engineering Consulting Corporation

Fig. 1.44 "China Star" Train

completed the special evaluation of the *Pre-feasibility Study Report on the Beijing–Shanghai High-speed Railway*, and completed the *Pre-feasibility Study Report on the Beijing–Shanghai High-speed Railway (Supplementary Draft for Evaluation)* based on the evaluation. In 2002, in order to implement the requirements of "*Notice on Reserving the Construction Land for Beijing–Shanghai High-speed Railway*" (JJC[2001]No. 2470) issued by State Planning Commission and Ministry of Land and Resources, additional survey was conducted along the whole line of Beijing–Shanghai High-speed Railway. The suggestion of local governments along the line were collected, and drawings of land use plan were submitted to local governments along the line. A special study for the whole line route was organized and communications were further conducted with local governments or authorities, and the *Feasibility Study Report for Beijing–Shanghai High-speed Railway* (Draft for Intermediate Review) was completed.

On January 7, 2004, *Medium and Long-term Railway Network Plan* was discussed and adopted in principle at the executive meeting of the State Council which put an end to the planning stage of high-speed railways and marked the beginning of the stage of large-scale implementation of construction of high-speed railways in China. The suggestions of the present experts of adopting wheel-track technology instead of magnetic levitation technique for Beijing–Shanghai High-speed Railway were accepted at the meeting.

(2) Implementation Stage

The ballastless track test section design and engineering test of Suining–Chongqing Railway was launched in September 2004, marking the beginning of the implementation stage of the high-speed railway and railway passenger dedicated line in China, which lasts up to now (see Fig. 1.45).

During this period, many high-speed railways and passenger dedicated lines including Beijing–Shanghai High-speed Railway, Wuhan–Guangzhou Passenger Dedicated Line, Zhengzhou-Xi'an Passenger Dedicated Line and Beijing–Tianjin Intercity Rail Transit Project were constructed successively, and high-speed railways were also built on the loess area, cold area, coastal areas and mountain-crossing areas, and the longest high-speed railway worldwide was built (see Fig. 1.46).

During the large-scale construction of high-speed railways, besides the ballastless track test section design and engineering test

Fig. 1.45 The First Ballastless Track Test Line of China–Suining–Chongqing Ballastless Track Test Line

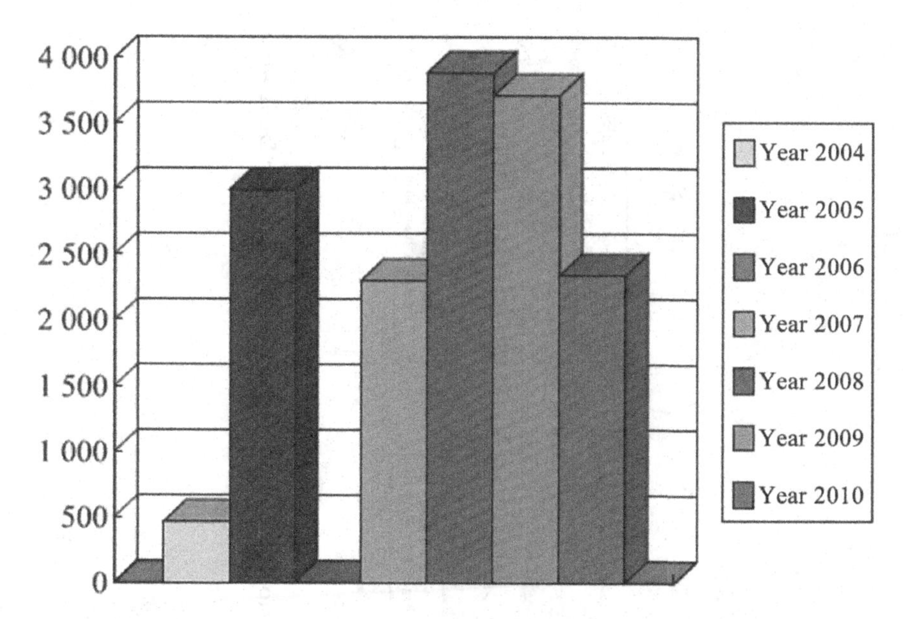

Fig. 1.46 Length of China's High-speed Railways Commenced from 2004 to 2010 (Unit: km)

of Suining–Chongqing Railway, the laying and running tests for the further technological innovation of ballastless track were conducted along a 62 km-long comprehensive test section of Wuhan–Guangzhou Passenger Dedicated Line (see Fig. 1.47).

(3) Realization stage

In April 2008, the operation of Hefei–Nanjing Passenger Dedicated Line marked the stepping into realization stage of China's high-speed railways. Beijing–Tianjin Intercity Rail Transit Project, Wuhan–Guangzhou Passenger Dedicated Railway, Zhengzhou–Xi'an Passenger Dedicated Railway, Beijing–Shanghai High-speed Railway have been put into operation successively.

The China-made "Harmony" CRH_3 EMU recorded a test speed of 394 km/h during the test run on Beijing–Tianjin Intercity Railway on June 24, 2008 and on Wuhan–Guangzhou Passenger Dedicated Line on December 9, 2009 and on Zhengzhou–Xi'an Passenger Dedicated Line on December 11, 2009 respectively. On September 28, 2010, the China-made "Harmony" CRH_{380A} EMU recorded a

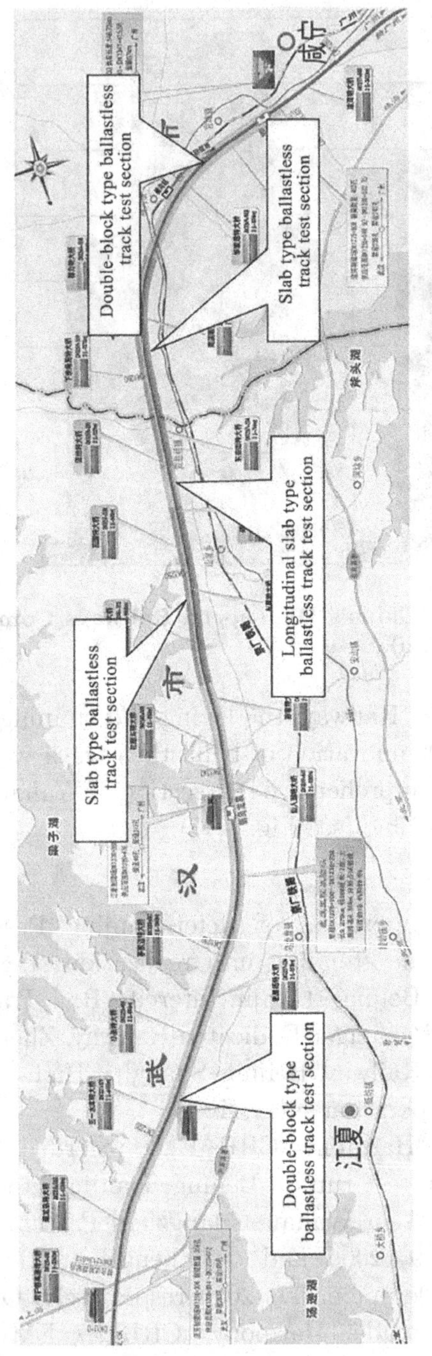

Fig. 1.47 Plan of Comprehensive Ballastless Track Test Section of Wuhan–Guangdong Passenger Dedicated Line

Fig. 1.48 A High-speed Train of China

test speed of 416.6 km/h during the trial run on Shanghai–Hangzhou Passenger Dedicated Line. On December 3, 2010, the China-made "Harmony" CRH_{380A} EMU recorded a test speed of 486.1 km/h during the joint commissioning of Beijing–Shanghai High-speed Railway (see Fig. 1.48).

Chapter 2

Management of Railway Engineering Consultation

2.1 Construction Process and Management of China's Railways

The construction process flow of China's railways is divided into the project initiation and decision-making stage, the project implementation stage and the project completion stage. Each stage consists of corresponding tasks, as shown in Fig. 2.1.

In the process flow mentioned above, there are four steps involving review related to design in the project implementation stage. According to *Measures for Survey and Design Management of Railway Construction Projects*, review system is applied to preliminary design documents and the review is organized by the Ministry of Railways. Review system is applied to construction drawings, and the review work is organized by the employer and supervised by the Ministry of Railways. According to *Regulations on the Quality Management of Railway Construction Projects*, the employer is responsible for conducting the preliminary review of the preliminary design and Class I design changes, approving Class II design changes, and organizing the consultation and review of construction drawings in compliance with the construction drawings. According to *Measures for Construction*

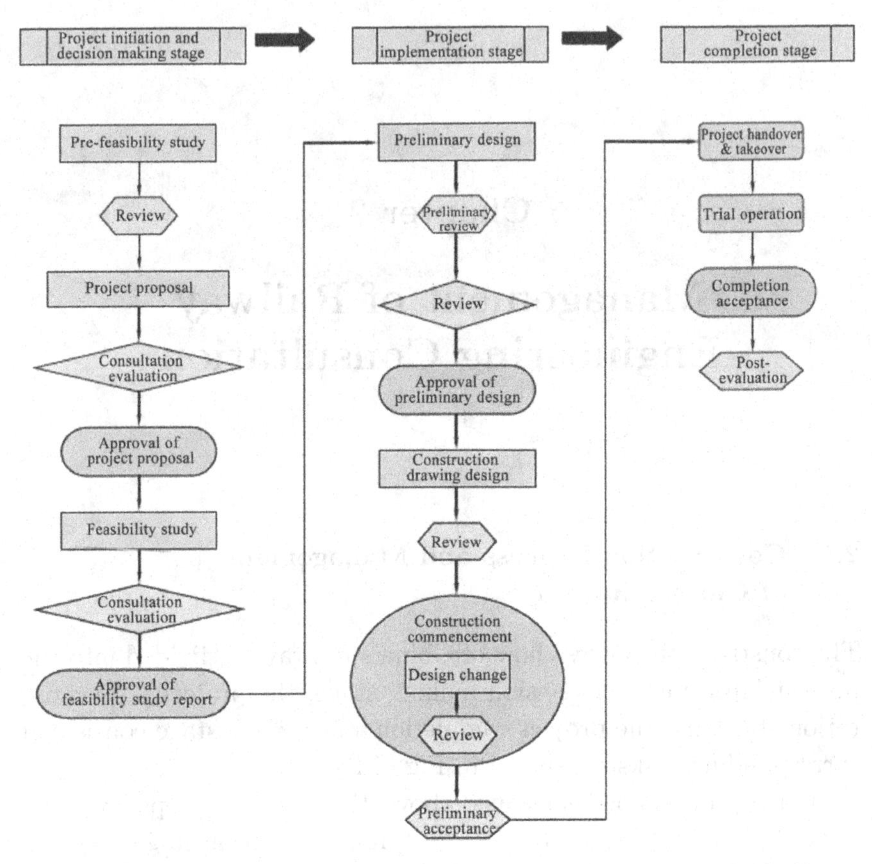

Fig. 2.1 Construction Process Flow of China's Large and Medium-scale Railway Projects

Drawing Review Management of Railway Construction Projects, the employer is responsible for the construction drawing review of construction projects. The implementation of specific review work may be entrusted to a company with Grade A qualification for railway design or railway engineering consultation (supervision). Moreover, it is clearly specified that the construction drawing review work is under centralized management of the Ministry of Railways. The employer may also entrust the preliminary review of preliminary design and Class I design changes to a consulting agency.

2.2 Organization Structure for Railway Engineering Consultation and Responsibilities of Key Personnel

Railway engineering consultation mainly includes design optimization consultation, construction drawing review, construction process consultation and special consultation. Generally, railway engineering consultation means the consultation process in which the employer, according to relevant regulations of the Ministry of Railways entrusts the consultation work to an agency with corresponding qualification for railway design or engineering consultation (supervision) (hereinafter referred to as the "consulting agency") through invitation for bids and the consulting agency carries out the consultation work in the form of setting up a project consultation department. Project manager responsibility system is applied to the project consultation department, and key personnel generally include project manager, chief consultant, discipline consultation director, planning and management director, as well as project risk management director if necessary.

2.2.1 *Project manager*

The project manager is the fully authorized principal of the consulting agency to fully perform the contract and is responsible for presiding over the management of project under consultation, organizing, coordinating and handling the internal and external relationships, and guaranteeing the achievement of planned objectives according to the requirements of the contract and relevant railway engineering consultation.

The project manager can represent the consulting agency to coordinate and communicate relevant issues with the employer, design institute and other relevant departments, as well as determine the consultation organization structure and clearly specify the work division of the chief consultant, the discipline consultation director, and the planning and management director.

The scope of responsibilities of the project manager also includes the following:

(1) Formulate the working policies for consultation and prepare the relevant working system for consultation. Direct the preparation of engineering consultation planning and risk management control documents and organize the implementation. Direct the establishment of a management system and its information system and take charge of the communication work for project management and its information.
(2) Supervise the work of the chief consultant and all departments. Preside over the management of the consultation contract and settle contractual disputes and claims related to the consulting agency. Organize the summary of consultation work.
(3) For a risk project requiring treatment from a consulting agency, determine the risk treatment principal based on risk assessment results, follow up on the project risk record forms provided by the principal, and get information on and follow up on the progress and result of risk treatment on relevant routine meetings. Provide and implement proper and feasible risk treatment measures including risk control and reduction measures and risk response plans under the assistance of risk management execution personnel. Coordinate with the project risk management execution personnel by referring to project risk record forms, review and approve risk assessment results, and provide professional consultation suggestions for the employer. Review and approve the risk management control documents including project risk assessment criteria and submit such documents to the employer. Promote the execution of relevant risk management work in compliance with management procedures and consultation progress.

2.2.2 *Chief consultant*

The chief consultant is the overall technical and quality director sent by the consulting agency to the project consultation department to fully perform the contract. According to the authorization by the project manager, the chief consultant presides over the technical and quality management work for railway engineering consultation and undertakes the overall responsibilities of consultation technology, quality and report. The chief consultant is responsible for presiding over the consultation work under the leadership of the project manager.

The scope of responsibilities of the chief consultant also includes the following:

(1) Assist the project manager in organizing the preparation and implementation of consultation plans. Organize the preparation of detailed rules for consultation implementation. Preside over the review of design principles, design schemes, monographic study reports and consultation reports.
(2) Preside over various comprehensive joint review, sign off on the comments, direct necessary trail checkout and sign consultation reports. Preside over the evaluation and analysis of major technical proposals and organize the coordination meeting for major construction organization plans. Review the consultation comments and suggestions on major technical proposals and technical issues, such as project scale, design standards and function requirements. Review the consultation comments and conclusions on whether the project meets the requirements for construction objectives in the aspects of safety, quality, construction period and investment. Review interdisciplinary interfaces and system integration documents.
(3) Preside over the summary of consultation work and review consultation reports, monthly reports and exception reports.

2.2.3 *Discipline consultation director*

The discipline consultation director is responsible for the consultation work of his/her own discipline and subject to the leadership and coordination of the chief consultant.

The scope of responsibilities of the discipline consultation director also includes the following.

(1) Prepare the implementation plans and detailed rules for disciplinary consultation and assist the chief consultant in preparing

and implementing detailed rules for consultation. Organize the disciplinary consultation work, participate in comprehensive joint reviews of technical documents and proposals, and make the disciplinary comments.

(2) Preside over disciplinary scheme review meetings and participate in disciplinary technical exchanges and negotiations. Review the interfaces and system integration documents related to his/her own discipline. Organize the inspection on the construction organization design and construction technology of his/her own discipline. Participate in the joint review on overall construction organization design and make the disciplinary comments. Propose the implementation plans, construction requirements, quality inspection means and product certification requirements for the application of new processes, new materials, new technologies and new equipment.

(3) Assist the chief consultant in conducting the management of consultation contracts and provide relevant data in time. Complete the summary of disciplinary consultation work in stages and sign disciplinary consultation reports and relevant documents. Keep records of consultation logs, collect the data and information related to the discipline and incorporate such data and information into the information management system, and assist with the management and archiving work of documents and data of his/her own discipline.

2.2.4 *Planning and management director*

The planning and management director is responsible for plan preparation, daily management and work coordination for consultation.

The scope of responsibilities of the planning and management director also includes the following.

(1) Compile and develop project management rules and regulations. Collect and sort consultation summaries and project management reports. Take charge of planning and scheduling, management of office equipment and facilities, and recording

of daily work. Master dynamic conditions of safety and quality work, take effective measures for existing problems and promote smooth project implementation.

(2) Prepare the consultation progress management plan, control progress plan and detailed progress plan, coordinate and solve the problems in the execution of progress plans, and take charge of the implementation and monitoring of progress plans. Check, understand and analyze the execution of progress plans, predict the potential factors affecting the progress of works, and provide relevant measures and suggestions.

(3) Assist the project manager in solving the problems found in the implementation of progress plans and facilitate the implementation of overall control progress plan. Organize and arrange for the training for consultation personnel. Implement the building of information management system and realize information system management.

2.2.5 *Project risk management director*

The project risk management director is responsible for preparing risk management control documents, developing and implementing rational risk management procedures, supervising the risk management work and proposing risk assessment criteria.

The scope of responsibilities of the project risk management director also includes the following.

(1) Execute the risk management work, preside over risk assessment and follow up meetings, and prepare project risk record forms. Designate the risk treatment principal for each risk according to risk assessment results and submit such risk treatment principals to the chief consultant for approval.

(2) Regularly update and report to the chief consultant the risk record forms according to the progress and results of risk treatment, and take note of relevant personnel in case of any risk management matter requiring timely follow up. Monitor risks and abnormalities in the consultation process. Provide the

employer with professional consultation comments reviewed by the chief consultant by referring to the project risk record forms.

2.3 Consultation Quality Plan

Establish consultation quality objectives according to the quality, schedule and investment control requirements for railway projects and railway engineering consultation as specified in the consultation contract signed with the employer.

Determine the consulting agency and the consultants, provide qualified consultation and management personnel and clearly specify their responsibilities and authorities in consideration of characteristics of the project and the consultation work, as well as communicate with the employer and interested parties in the form of documents.

Prepare the consultation quality plan according to the requirements of ISO standards, current national and industrial laws and regulations, technical policies, approval documents and technical standards for the project, consultation contract and standardization management.

Propose quality control standards and index systems and prepare detailed rules for consultation implementation according to the scope, requirements and quality objectives of consultation work for each discipline.

The consultation quality plan mainly includes the following (see Fig. 2.2).

(1) Project profile, scope and basis;
(2) Organization structure and personnel of the project consultation department, and their responsibilities, authorities and work division;
(3) Overall objectives of consultation quality and consultation quality objectives of each discipline;
(4) Scope, method and quality control measures for consultation, including the time and methods for verification and review of disciplinary technical requirements and consultation documents;
(5) Input, information and facilities of consultation;

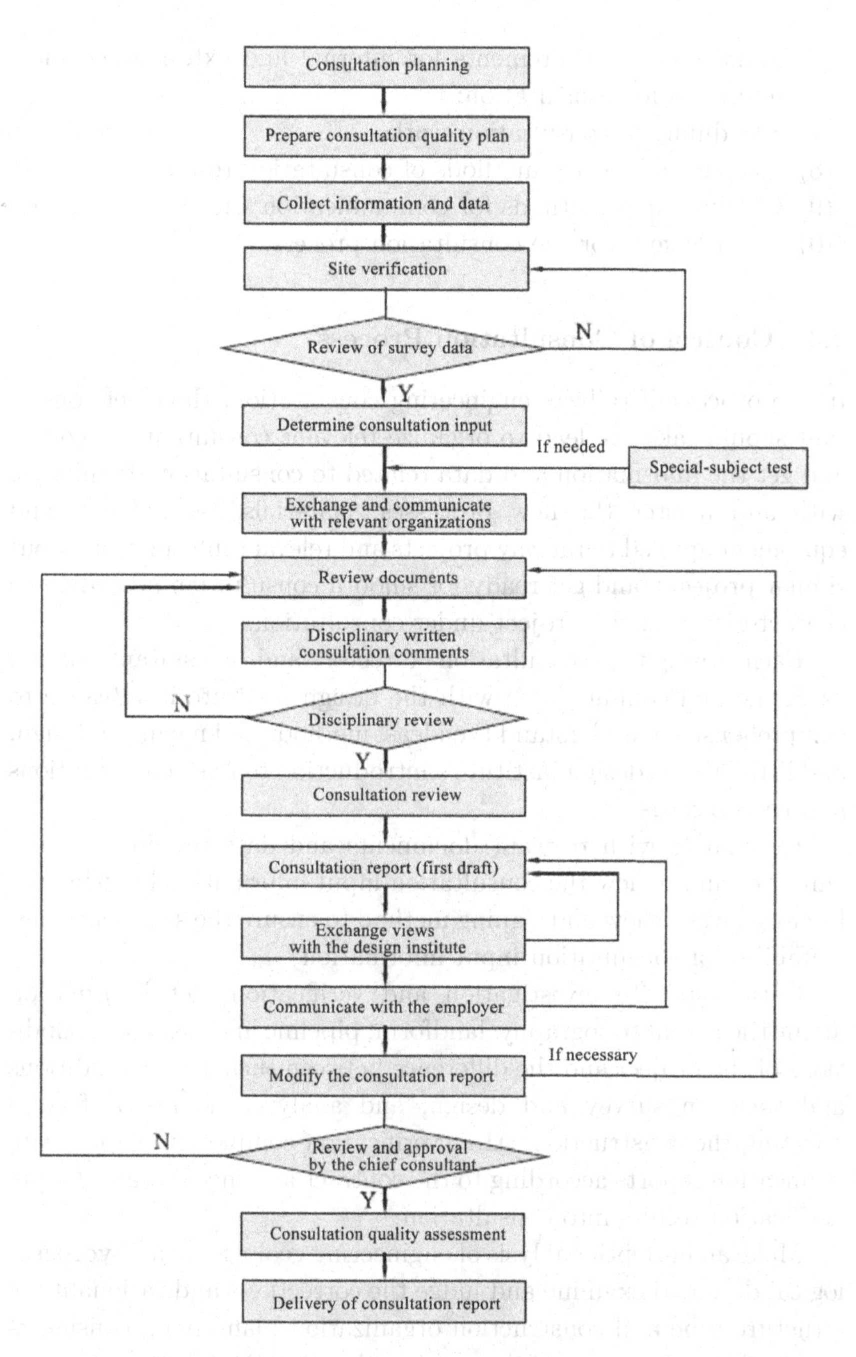

Fig. 2.2 Quality Control of Engineering Consultation

(6) Management requirements for internal and external technical interfaces for consultation;
(7) Scheduling of consultation work;
(8) Output and review methods of consultation results;
(9) Channels and methods for communication with the employer;
(10) Records kept for the consultation process.

2.4 Control of Consultation Process

In the process of railway engineering consultation, the chief consultant should take the lead to organize relevant consultants to collect and get the information and data related to consultation, familiarize with and master the new processes, materials, technologies and equipment applied to railway projects and relevant information about similar projects, and get ready for smooth consultation according to characteristics of the project under consultation.

Each discipline consultation director and consultant should exchange and communicate with the design institute in advance to comprehensively understand the ideas, methods and means of design, and listen to the design institute's introduction to relevant conditions and keep records.

Familiarize with relevant documents and data provided by the employer and review the consultation input information by using the level-by-level review and signing method to ensure the sufficiency and suitability of consultation input information.

Carry out site investigation and verification, get information about the actual topography, landform, pipeline and transport conditions of the project and the difference between such actual conditions and those in survey and design, and analyze the major factors affecting the construction. After verification completion, submit site verification reports according to the contract and incorporate the site verification results into consultation.

Make an in-depth analysis of engineering geological and hydrogeological data and examine and judge the correctness and rationality of structure type and construction organization plan giving consideration to the environment of the construction site. Review the interfaces for systematic design of civil and E&M works according to the design and construction experience of similar projects.

Review the basic data such as technical requirements, document compositions and contents and unified provisions for design to control the correctness of design input and the rationality and uniformity of technical standards and document depth from the beginning. Comprehensively and correctly implement relevant approved suggestions and review the standards, requirements and modes of operation functions.

Each discipline consultation director should give consultation comments on design documents by using such consultation methods as site test, model test, checking calculation analysis and comparison analysis according to the consultation quality plan.

The discipline consultation director should propose and the chief consultant should lead, in the form of meeting, the review of technical issues involving multiple disciplines, major technical proposals, technical difficulties, systematic technical issues and major technical issues about the application of new processes, materials, technologies and equipment to identify the existing problems and propose necessary measures.

Pay attention to the review of engineering system design, technical standards, service functions, overall design principles, discipline design principles and detailed rules, discipline interfaces, depth and breadth of survey and design, control works, bill of quantities and investment, major construction technical proposals, construction processes, and standard, general and reference drawings with broad influence.

Review system compositions, functional standards and type selection of equipment and coordinate the relation between the system and civil works to achieve the objectives for coordinative matching.

Apply the induction, summary, predictive analysis and analogy analysis and deduction methods, propose optimized, improved and modified consultation comments or conclusions and fill out consultation record forms.

The discipline consultation director should review the consultation comments or conclusions of his/her own discipline, verify the integrity and correctness of consultation, propose the consultation review results and necessary improvement measures, and fill out consultation review forms.

Exchange and communicate with the centralized management department of the Ministry of Railways and the employer by meeting, letters and phone to find out the requirements for consultation work, listen to the comments and suggestions on consultation results and continuously improve the quality of consultation service.

2.5 Control of Consultation Results

2.5.1 *Preparation of consultation reports*

Prepare the consultation report in the principles of sufficient basis, accurate analysis, overall evaluation, outstanding theme, clear conclusion and strong pertinence and technical predictability.

Regularly analyze design quality, progress and investment control conditions, propose consultation comments and conclusions, and prepare monthly review reports. The discipline consultation director should summarize the consultation comments and conclusions of his/her own discipline, communicate and coordinate with relevant disciplines, compile the regulations and relevant requirements of the consultation report, and prepare the disciplinary consultation report.

The chief consultant should organize the technical review personnel or experts to review the disciplinary consultation reports and propose review conclusions and modification suggestions, the modifications of which should be submitted to the discipline consultation director for review and to the chief consultant for approval before forming the engineering consultation report.

Relevant consultation documents such as disciplinary consultation comments and consultation reports should be delivered to the employer after verification, review, disciplinary countersigning and approval according to the signed provisions in consultation documents.

2.5.2 *Assessment and modification of consultation results*

The discipline consultation director should carry out daily supervision and inspection on the consultation process and propose improvement comments and suggestions on the problems that have arisen according to the quality objectives and requirements determined in consultation planning.

Before the delivery of the design consultation report, the chief consultant should organize relevant personnel to carry out quality assessment of the consultation report, make modifications to the problems identified and consider the quality assessment result as the assessment basis for consultation quality.

After the delivery of consultation report, the employer and external experts should be invited to review the consultation results and sign the review conclusions.

2.5.3 *Control of unqualified consultation service*

In the consultation process, it is a requirement to collect, know about and consult the comments of the centralized management department of the Ministry of Railways and the employer on consultation work, irregularly organize the review, make treatment proposals and rectify according to review comments. Unqualified consultation results should be identified. The chief consultant should take the lead to organize the review, propose treatment suggestions and corrective and preventive measures. The relevant discipline consultation director should make rectifications according to the review comments. The conformity of consultation results should be analyzed regularly, corresponding improvement measures should be developed and consultation quality should be improved continuously.

2.6 Consultation Management Systems

Consultation management systems are divided into the administrative management systems and the technical management systems.

2.6.1 *Administrative management systems*

Administrative management systems mainly include regulations on job responsibilities, regulations on labor disciplines, assessment and distribution methods, financial management method, regulations on the management of work safety, regulations on self-discipline and anti-corruption, management method for operation planning and scheduling, examination and evaluation system, and warning system for violation of disciplines and regulations, and dereliction of duty.

(1) The regulations on job responsibilities mainly specify the job responsibilities and work division of the members at all levels of the consulting agency.
(2) The regulations on labor disciplines mainly specify the methods, procedures and standards for checks on work attendance, leave applications, personnel appraisals, praise and punishment in the consulting agency.
(3) The assessment and distribution methods mainly formulate the assessment methods for all departments and personnel of the consulting agency, including the scope, type, personnel, procedures, standards and contents of assessment, as well as the regulations on distribution, praise and punishment.
(4) The financial management method mainly stipulates the specific method for financial management in the consulting agency and clearly specify the scope of reimbursement, the provisions on original documents (reimbursement documents), the capital management method and the management of fixed assets and low-value consumables.
(5) The regulations on the management of work safety mainly specify the safe behaviors of all departments and members of the consulting agency in their work and life.
(6) The regulations on self-discipline and anti-corruption mainly specify the requirements for all departments and members of the consulting agency in their work and life during consultation with regard to self-discipline and anti-corruption.

(7) The management method for operation planning and scheduling mainly specifies the plans, schedules, working procedures, etc. for consultation work.

(8) The examination and evaluation system mainly examines and evaluates whether during the consultation the contract is performed accurately whether all kinds of work arranged by the employer are finished satisfactorily, whether the consultation work is carried out in accordance with the prepared consultation quality plan and the detailed rules for consultation implementation, whether various types of inspection certificate meet the specified requirements, whether all accounts and management systems are established and implemented as required, and whether the consultation personnel have any behaviors violating the national laws, regulations and relevant rules of the project consultation department with regard to self-discipline and anti-corruption, as well as the employer's comments on consultation quality and the effect of consultation on construction quality, progress, investment and safe and civilized operation.

(9) The warning system for violation of disciplines and regulations and dereliction of duty mainly inspects and assesses the examination and assessment rules at a regular interval, and punishes and circulates the behaviors such as non-performance of duties, providing consultation service not in compliance with specified procedures, perfunctory performance, fraud, injury to goodwill of the consulting agency, as well as the behaviors that lead to construction quality accidents due to negligence or dereliction of duty and faults, aim at seeking private interests illegally and result in economic loss.

2.6.2 *Technical management systems*

Technical management systems mainly include regulations on technical management, periodic report system, work report system, signing regulations, regulations on document management, regulations on archiving management and meeting system.

(1) The regulations on technical management mainly specify the purpose of consultation and stipulate the consultation bases, principles, overall requirements, key points, working procedures and methods, requirements for format and content of consultation documents, registration format of communication documents, responsibilities of consulting agency and consultation personnel.
(2) The periodic report system mainly handles the main contents, methods and procedures for preparing and reviewing monthly, quarterly and annual reports with regard to consultation, as well as the submission time and feedback information of such reports.
(3) The work report system mainly specifies the procedures and types of reports concerning major matters such as safety, quality, progress, investment and environmental protection, the stage division and submission of single-subject consultation report, and the procedures, formats and requirements for submission of the summary consultation report.
(4) The signing regulations mainly specify the main contents and disciplines of the consultation report to be reviewed and the reviewing and signing requirements for the consultation report involving multiple disciplines when the consultation report is signed.
(5) The regulations on document management mainly specify the preparation, review, approval, issuance, modification and identification requirements for and the regulations on receiving, delivering, using and keeping the documents related to consultation.
(6) The regulations on archiving management mainly specify the personnel, work contents and requirements for the collection, sorting, archiving and management of consultation data, documents and reports.
(7) The meeting system mainly establishes the system of regular or irregular disciplinary coordination meetings, technical review meetings (including expert review meetings), operation scheduling meetings, work shift meetings, as well as weekly (10-day) reporting meetings, monthly work planning and summing-up meetings and workshops.

2.7 Working Procedures of Consultation

2.7.1 *Design optimization consultation (see Fig. 2.3)*

(1) Various preparations for performing design optimization work should be made according to the requirements of the consultation contract signed by and between the consulting agency and the employer, the principles and detailed rules for design optimization.

(2) The consulting agency should make the design optimization stage plan, progress control plan and human resources plan for the project based on the delivery time for design documents

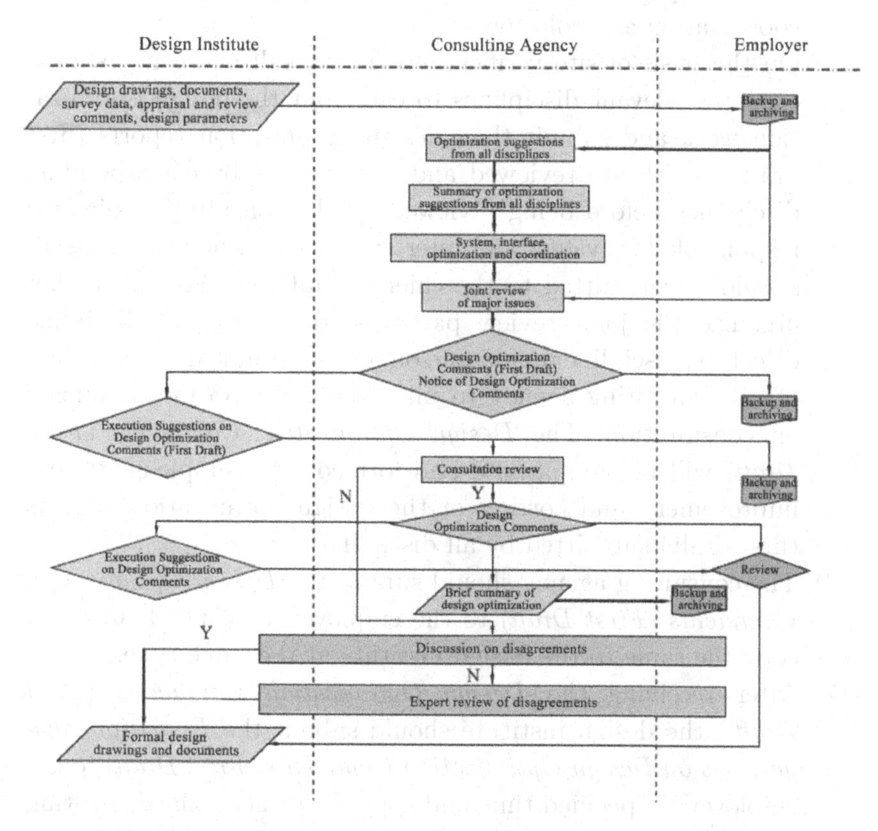

Fig. 2.3 Process Flow Chart of Design Optimization Consultation

and data provided by the employer, and rationally assign the design optimization personnel to meet the progress and quality requirements of design optimization.

(3) The employer should provide the consulting agency with relevant data necessary for design optimization, including various drawings, documents, review and approval comments, in accordance with the determined schedule.

(4) After receiving the design documents and data provided by the employer, the consulting agency should register and number the design documents and data as required. Each discipline is responsible for sorting the design documents and data received. If the design documents and data are incomplete, a written report should be submitted to the employer in time for coordination and solution.

(5) On the basis of site verification, the consulting agency should organize relevant disciplines to carry out the design optimization work and submit their design optimization reports (first draft), which are reviewed and summarized by corresponding disciplines before being reviewed by the consultation director responsible for system and interface consultation. Major issues should be submitted to the chief consultant, who will further organize the joint review participated by relevant discipline directors, discipline review personnel and technical expert group of the consulting agency to make the joint review comments on consultation. The *Design Optimization Comments* (*First Draft*) will be summarized and formed after supplementation, improvement and sorting of the design optimization reports (first draft) submitted by all disciplines.

(6) The consulting agency should submit the *Design Optimization Comments* (*First Draft*) to the employer in planned time and copy the same to the design institute at the same time.

(7) After receiving the *Design Optimization Comments* (*First Draft*), the design institute should submit the *Execution Suggestions on Design Optimization Comments* (*First Draft*) to the employer in specified time and copy the same to the consulting agency at the same time.

(8) After receiving the *Execution Suggestions on Design Optimization Comments* (*First Draft*) from the design institute, the consulting agency should immediately organize all disciplines to analyze and study the *Execution Suggestions on Design Optimization Comments* (*First Draft*), communicate and exchange with the design institute in time, form the *Review Report* of *Design Optimization* and submit it to the employer, and copy the same to the design institute at the same time.

(9) When the design institute has no objection to the *Design Optimization Comments* (*First Draft*) or reaches a consensus after communication and exchange, the consulting agency and the design institute will submit the *Design Optimization Comments* (*First Draft*) and the *Execution Suggestions on Design Optimization Comments* (*First Draft*) as formal *Design Optimization Comments* and *Execution Suggestions on Design Optimization Comments* to the employer for review and approval.

(10) The design institute should carry out the design optimization work according to the *Design Optimization Comments* approved by the employer, and the employer should provide the consulting agency with the *Design Optimization Comments* in a timely manner, which serves as one of the important documents confirmed for design optimization.

(11) When the consulting agency has objections to the suggestions of the design institute and fails to reach a consensus after communication, this condition should be reported to the employer. After that, the chief consultant of the consulting agency should preside over a technical coordination meeting participated by the employer and the design institute. "Minutes of Meeting" should be kept at the meeting and submitted to the design institute for execution. If no consensus can be reached by doing so, the employer should be notified. If necessary, the chief consultant of the consulting agency should preside over an expert review meeting attended by technical leaders and experts of the employer and the design institute to carry out special-subject review and form the final comments which will be submitted to the design institute for execution.

2.7.2 *Construction drawing review consultation (see Fig. 2.4)*

(1) Various preparations for performing construction drawing review work should be made according to the requirements of the consultation contract signed by and between the consulting agency and the employer, the principles and detailed rules for construction drawing review.

(2) The consulting agency should make the construction drawing review stage plan, progress control plan and human resources plan for the project based on the delivery time for construction drawings, documents and data provided by the employer, and rationally assign the construction drawing review personnel to

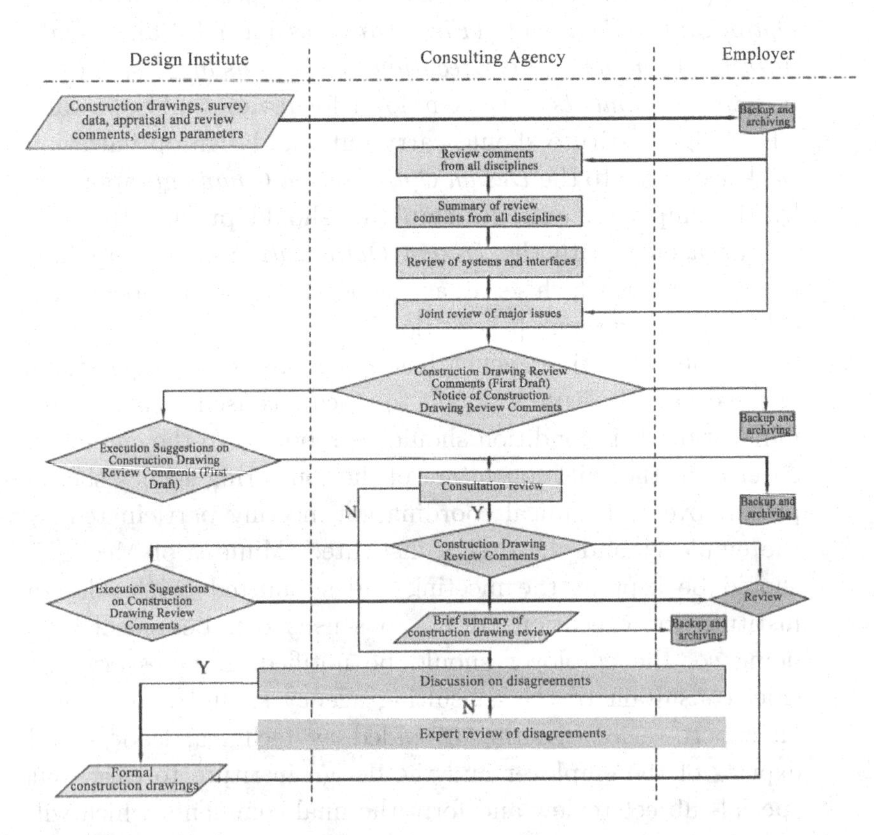

Fig. 2.4 Process Flow Chart of Construction Drawing Review

meet the progress and quality requirements of construction drawing review.

(3) The employer should provide the consulting agency with relevant data necessary for construction drawing review, including various drawings, documents, review and approval comments at all levels, in accordance with the determined schedule.

(4) After receiving the construction drawings, documents and data provided by the employer, the consulting agency should register and number the construction drawings, documents and data as required. Each discipline is responsible for sorting the construction drawings, documents and data received. If the construction drawings, documents and data are incomplete, a written report should be submitted to the employer in time for coordination and solution.

(5) On the basis of site verification, the consulting agency should organize relevant disciplines to carry out the construction drawing review work and submit their construction drawing review reports (first draft), which are reviewed and summarized by corresponding disciplines before being reviewed by the consultation director responsible for system and interface consultation. Major issues should be submitted to the chief consultant, who will further organize the joint review to be participated by relevant discipline directors, discipline review personnel and technical expert group of the consulting agency to make the joint review comments on consultation. The *Construction Drawing Review Comments* (*First Draft*) will be summarized and formed after supplementation, improvement and sorting of the construction drawing review reports (first draft) submitted by all disciplines.

(6) The consulting agency should submit the *Construction Drawing Review Comments* (*First Draft*) to the employer in time and copy the same to the design institute at the same time.

(7) After receiving the *Construction Drawing Review Comments* (*First Draft*), the design institute should change its design in accordance with the *Construction Drawing Review Comments* (*First Draft*), finish the construction drawing modification

work in specified time, submit the *Execution Suggestions on Construction Drawing Review Comments* (*First Draft*) to the employer, and copy the modified construction drawings and *Execution Suggestions on Construction Drawing Review Comments* (*First Draft*) to the consulting agency.

(8) After receiving the *Execution Suggestions on Construction Drawing Review Comments* (*First Draft*) from the design institute, the consulting agency should immediately organize all disciplines to review and confirm the modified construction drawings and analyze and study the *Execution Suggestions on Construction Drawing Review Comments* (*First Draft*), communicate and exchange with the design institute in time, form the *Review Report* of *Construction Drawing Review* and a notice about the review and submit them to the employer, and copy the same to the design institute at the same time.

(9) When the design institute has no objection to the *Construction Drawing Review Comments* (*First Draft*) or reaches a consensus after communication and exchange and the construction drawings have been reviewed and confirmed by the director of construction drawing review discipline, the consulting agency and the design institute will submit the *Construction Drawing Review Comments* (*First Draft*) and the *Execution Suggestions on Construction Drawing Review Comments* (*First Draft*) as formal *Construction Drawing Review Comments* and *Execution Suggestions on Construction Drawing Review Comments* to the employer. The employer will provide the consulting agency with the design institute's *Execution Suggestions on Construction Drawing Review Comments*, which serves as one of the important documents confirmed for construction drawing design in construction drawing review, and issue a notice on Printing Construction Drawings to the design institute to print formal construction drawings.

(10) When the consulting agency has objections to the suggestions of the design institute and fails to reach a consensus after communication, this condition should be reported to the employer. After that, the chief consultant of the consulting agency should

preside over a technical coordination meeting attended by the employer and the design institute. "Minutes of Meeting" should be kept at the meeting and submitted to the design institute for execution. If no consensus could be reached by doing so, the employer should be notified. The chief consultant of the consulting agency should preside over an expert review meeting attended by technical leaders and experts of the employer and the design institute to carry out subject-special review and form the final comments which will be submitted to the design institute for execution.

2.7.3 *Construction process consultation (see Fig. 2.5)*

(1) Various preparations for performing construction process consultation work should be made according to the requirements of the consultation contract signed by and between the consulting agency and the employer, the principles and detailed rules for construction process consultation.
(2) The consulting agency should make the construction process consultation stage plan, progress control plan and human resources plan for the project based on the site verification and in consideration of the issues or changed design documents and data provided by the employer in the construction process, and rationally assign the design optimization personnel to meet the progress and quality requirements of construction process consultation.
(3) The employer should provide the consulting agency with the relevant data necessary for consultation (including site engineering conditions, construction equipment and construction organization), the minutes of meetings and the review and approval comments of relevant parties necessary for review of design changes, in accordance with the determined schedule.
(4) After receiving the documents, drawings and data provided by the employer, the consulting agency should register and number the documents, drawings and data as required. Each discipline is responsible for sorting the documents, drawings and data

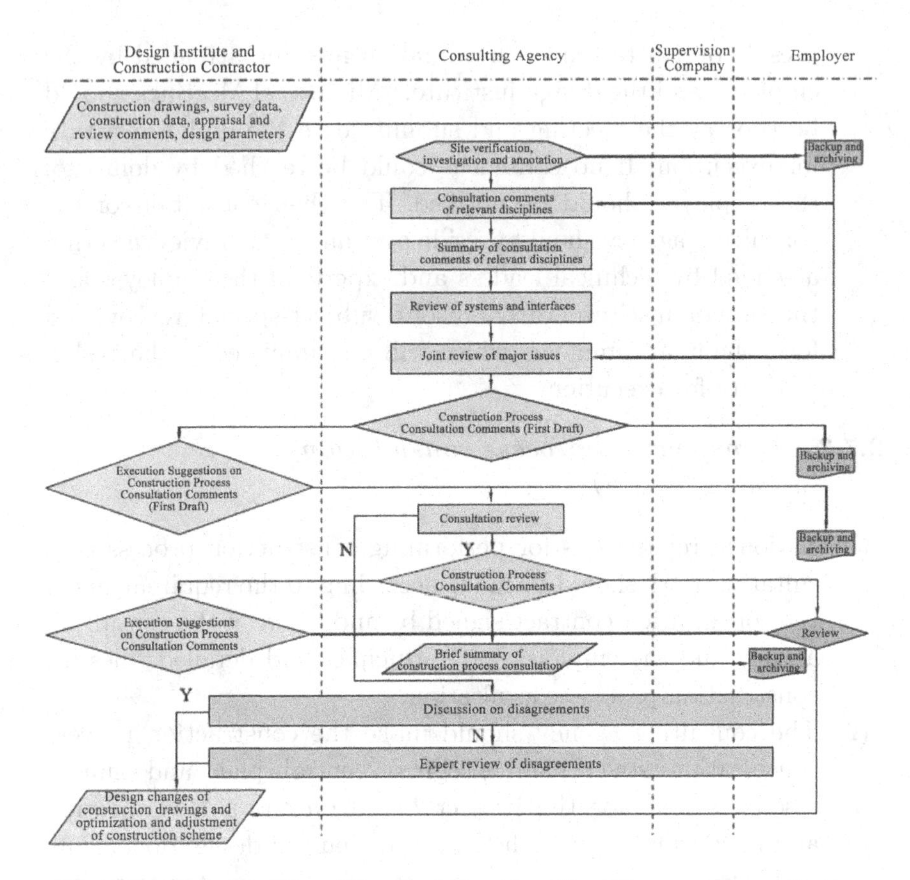

Fig. 2.5 Process Flow Chart of Construction Process Consultation

received. If the documents, drawings and data are incomplete, a written report should be submitted to the employer in time for coordination and solution.

(5) The consulting agency should organize relevant disciplines to carry out the construction process consultation work and submit their construction process consultation reports (first draft), which are reviewed and summarized by corresponding disciplines before being reviewed by the consultation director responsible for system and interface consultation. Major issues should be submitted to the chief consultant, who will further organize the joint review participated by relevant discipline

directors, discipline review personnel and technical expert group of the consulting agency to make the joint review comments. The *Construction Process Consultation Comments* (*First Draft*) will be formed after supplementation, improvement and sorting of the construction process consultation reports (first draft) submitted by relevant disciplines.

(6) The consulting agency should submit the *Construction Process Consultation Comments* (*First Draft*) to the employer in planned time and copy the same to the design institute and the construction contractor at the same time.

(7) After receiving the *Construction Process Consultation Comments* (*First Draft*), the design institute and the construction contractor should modify the design or construction scheme in accordance with the *Construction Process Consultation Comments* (*First Draft*), finish the design changes of construction drawings or the optimization and adjustment of construction scheme in specified time, submit the *Execution Suggestions on Construction Process Consultation Comments* (*First Draft*) to the employer, and copy the design changes of construction drawings or the optimization and adjustment of construction scheme and the *Execution Suggestions on Construction Process Consultation Comments* (*First Draft*) to the consulting agency.

(8) After receiving the *Execution Suggestions on Construction Process Consultation Comments* (*First Draft*), the consulting agency should immediately organize relevant disciplines to review and confirm the design changes of construction drawings or the optimization and adjustment of construction scheme and analyze and study the *Execution Suggestions on Construction Process Consultation Comments* (*First Draft*), communicate and exchange with the design institute or the construction contractor in time, form the Review Report and a notice on Design Changes of Construction Drawings and the optimization and adjustment of construction scheme, and copy the same to the design institute or the construction contractor at the same time.

(9) When the design institute or the construction contractor has no objection to the *Construction Process Consultation Comments* (*First Draft*) or reaches a consensus after communication and exchange and the design changes of construction drawings and the optimization and adjustment of construction scheme have been reviewed and confirmed by the discipline consultation director, the consulting agency and the design institute or the construction contractor will submit the *Construction Process Consultation Comments* (*First Draft*) and the *Execution Suggestions on Construction Process Consultation Comments* (*First Draft*) as formal *Construction Process Consultation Comments* and *Execution Suggestions on Construction Process Consultation Comments* to the employer. The employer will provide the consulting agency with the design institute's or the construction contractor's formal comments in a timely manner, which serves as one of the important documents confirmed for the design changes of construction drawings and the optimization and adjustment of construction scheme in construction process consultation, and issue a notice on Design Changes of Construction Drawings to the design institute to execute the formal design changes of construction drawings.

(10) When the consulting agency has objections to the suggestions of the design institute or the construction contractor and fails to reaches a consensus after communication, this condition should be reported to the employer. After that, the chief consultant of the consulting agency should preside over a technical coordination meeting attended by the employer, the design institute, the supervision company and the construction contractor. "Minutes of Meeting" should be kept at the meeting and submitted to the design institute or the construction contractor for execution. If no consensus could be reached by doing so, the employer should be notified. The chief consultant of the consulting agency should preside over an expert review meeting attended by technical leaders and experts of the employer, the design institute, the supervision company and the construction contractor to carry out subject-special review

and provide the final comments which will be submitted to the design institute or the construction contractor for execution.

2.7.4 *Special consultation (see Fig. 2.6)*

(1) Various preparations for performing special consultation work should be made according to the requirements of the consultation contract signed by and between the consulting agency and the employer, the principles and detailed rules for special consultation.

(2) Personnel of relevant disciplines should make the special consultation stage plan, progress control plan and human resources

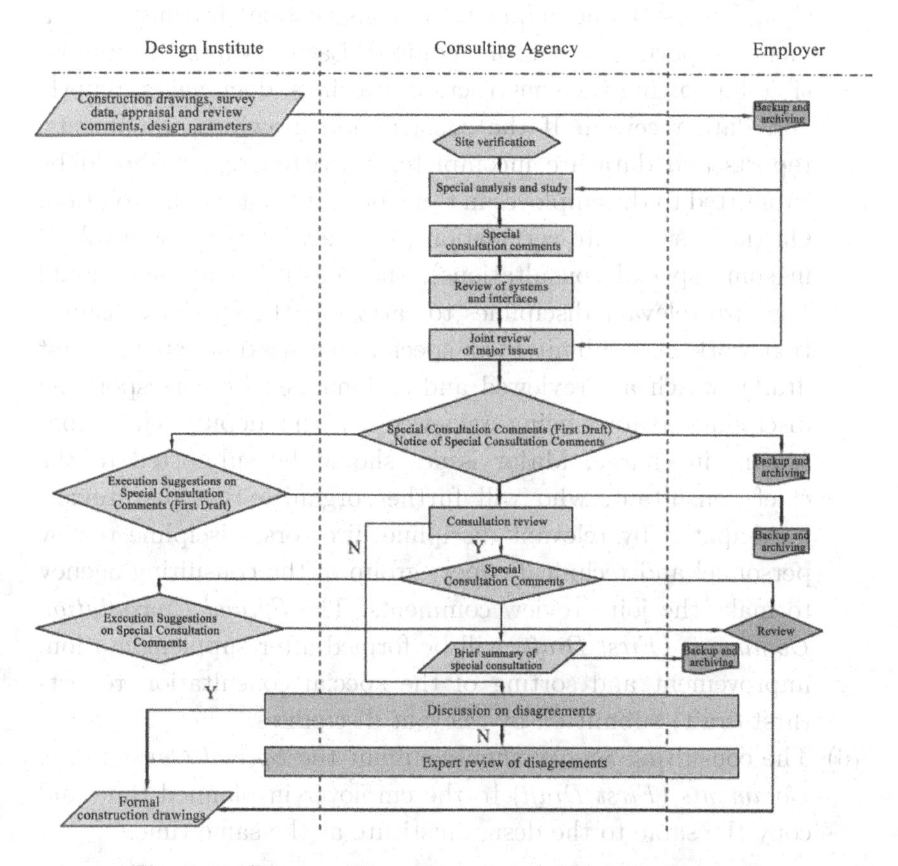

Fig. 2.6 Process Flow Chart of Special Consultation

plan for the project based on the delivery time for construction drawings, documents, reports and data provided by the employer. Special consultation personnel should be rationally assigned to meet the progress and quality requirements of special consultation.

(3) The employer should provide the consulting agency with the construction drawings, documents, reports and data necessary for special consultation, including various drawings, documents, review and approval comments at all levels, in accordance with the determined schedule.

(4) After receiving the construction drawings, documents, reports and data provided by the employer, the consulting agency should register and number the construction drawings, documents, reports and data as required. Each discipline is responsible for sorting the construction drawings, documents, reports and data received. If the construction drawings, documents, reports and data are incomplete, a written report should be submitted to the employer in time for coordination and solution.

(5) On the basis of site verification (no site verification is involved in some special consultations), the consulting agency should organize relevant disciplines to carry out the special consultation work and submit their special consultation reports (first draft), which are reviewed and summarized by corresponding disciplines before being reviewed by the deputy chief consultant in charge. Major issues should be submitted to the chief consultant, who will further organize the joint review participated by relevant discipline directors, discipline review personnel and technical expert group of the consulting agency to make the joint review comments. The *Special Consultation Comments* (*First Draft*) will be formed after supplementation, improvement and sorting of the special consultation reports (first draft) submitted by relevant disciplines.

(6) The consulting agency should submit the *Special Consultation Comments* (*First Draft*) to the employer in planned time and copy the same to the design institute at the same time.

(7) After receiving the *Special Consultation Comments* (*First Draft*), the design institute should change its design in accordance with the *Special Consultation Comments* (*First Draft*), finish the construction drawing modification work in specified time, submit the *Execution Suggestions on Special Consultation Comments* (*First Draft*) to the employer, and copy the modified construction drawings and *Execution Suggestions on Special Consultation Comments* (*First Draft*) to the consulting agency.
(8) After receiving the *Execution Suggestions on Special Consultation Comments* (*First Draft*) from the design institute, the consulting agency should immediately organize all disciplines to review and confirm the modified construction drawings and analyze and study the *Execution Suggestions on Special Consultation Comments* (*First Draft*), communicate and exchange with the design institute in time, form the *Review Report of Special Consultation* and submit it to the employer, and copy the same data to the design institute at the same time.
(9) When the design institute has no objections to the *Special Consultation Comments* (*First Draft*) or reaches a consensus after communication and exchange and the construction drawings have been reviewed and confirmed by the director of special consultation discipline, the consulting agency and the design institute will submit the *Special Consultation Comments* (*First Draft*) and the *Execution Suggestions on Special Consultation Comments* (*First Draft*) as formal *Special Consultation Comments* and *Execution Suggestions on Special Consultation Comments* to the employer. The employer will provide the consulting agency with the design institute's *Execution Suggestions on Special Consultation Comments*, which serves as one of the important documents confirmed for special consultation, and issue a notice on Printing Construction Drawings to the design institute to print formal construction drawings.
(10) When the consulting agency has objections to the suggestions of the design institute and fails to reach a consensus after communication, this condition should be reported to the

employer. After that, the chief consultant (or deputy chief consultant) of the consulting agency should preside over a technical coordination meeting attended by the employer and the design institute. "Minutes of Meeting" should be kept at the meeting and submitted to the design institute for execution. If no consensus can be reached by doing so, the employer should be notified. The chief consultant of the consulting agency should preside over an expert review meeting attended by technical leaders and experts of the employer and the design institute to carry out special-subject review and form the final comments which will be submitted to the design institute for execution.

2.8 Consultation Risk Management

For railway engineering consultation, scientific, systematic, complete and operable planning programs and management measures should be established and implemented and the consultation management level should be improved to achieve the optimal combination of consultation quality, construction period and cost. The consulting agency should establish an overall control plan for risk management with the "risk-based" project management concept and by referring to international and domestic risk management experience and relevant application standards.

Risk is defined as the possibility of loss, injury, disadvantage or damage. The risk of project under consultation means the possible effects on safety, quality, construction period and finance caused by uncertain factors in the aspects of decision-making, construction period, administration, finance and technology in the consultation and management processes. The main purpose of project risk management is to identify the potential risks of project activities and the possible abnormalities and their influences in the process of executing or managing these activities by following a set of systematic and structural management procedures, understand the risk-bearing capacity and ensure that the project can be successfully completed on schedule, at specified cost and according to the objectives by analyzing, managing, controlling and mitigating the project risks.

The main risks of railway engineering consultation are classified into internal risks and external risks. Internal risks mainly refer to whether the internal management mechanism of the consulting agency is sound and effective in operation, whether the work objectives are correct, the qualification of the consulting agency, the ability in handling the relations between the design institute and the employer and the external and internal relations, and the work and professional ethics accomplishment of engineering consultation personnel. External risks mainly refer to the degree of difficulty and ease in engineering technology, the matching degree between consultation duration and quality requirement, the adaptability of testing means, the reliability of analysis software and methods, the complexity of systems and interfaces, the changes of technical standards and total construction period, the quality of engineering survey and design, capital and cost, etc.

For risk management, it is necessary to identify the risk factors from the sources, rationally determine the risk levels, give up or modify high-risk conditions, propose the measures for risk treatment and monitoring, conduct systematic risk management, improve the risk management level, mitigate various potential risks in consultation to an acceptable level under the precondition of rationality and feasibility, guarantee safety and quality, control construction period and cost, increase benefits and protect the environment. Dynamic adjustment should be made in the implementation of risk management to ensure successful execution of risk management.

2.8.1 *Overall requirements*

Various risk factors in railway engineering consultation should be analyzed and recognized and risk management plans and programs should be prepared to cope with risks systematically. Assistance should be provided for the employer to identify critical factors and risks should be mitigated to the best effect by analysis, management and control, so as to ensure that the objectives for consultation quality, construction period and cost are controlled effectively.

An integrated risk management system, which is actively utilized by all consultation personnel and is for the comprehensive management of all risk factors, should be established. Risk factors should be tracked, managed, assessed, controlled and completed, and feedback should be given to the employer according to the risk management plan. Risk register and other means should be used to track and control risks in the whole process.

Consultation work should be improved and scientific, systematic, complete and operable consultation planning programs and management measures should be established and implemented. Risk management control documents should be prepared and rational procedures for risk management should be specified clearly.

2.8.2 *Basic procedures*

Basic procedures for risk management include preparation of risk management plan, risk assessment, risk treatment and risk monitoring. Among these procedures, risk assessment is the basis of risk management, and risk treatment and risk monitoring are post-treatment and application of risk assessment.

Preparation of risk management plan is mainly to determine the risk objectives, principles and strategies, propose the objective, scope, method and assessment standard for stage work, clearly specify the responsibilities of all participating parties, organize and carry out risk management and monitoring of all participating parties, and specify the contents and formats of relevant reports.

Risk assessment is to check, supervise, coordinate and deal with the issues related to assessment.

Risk treatment is mainly to review high and extremely high risk levels and make decisions on risk treatment measures.

Risk monitoring is mainly to prepare risk monitoring plan, propose monitoring standards and entrust a relevant professional institution to carry out risk inspection if necessary.

2.8.3 *Treatment and monitoring of risks*

There are four major measures for risk treatment, i.e. acceptance, mitigation, transfer and avoidance. Corresponding treatment measures should be taken according to risk assessment results and risk acceptance criteria to form a list of risk countermeasures. Low-level risks should be neglected, corresponding measures should be taken to mitigate high-level and medium-level risks, and extremely high-level risks should be avoided.

Whole-process monitoring of risk identification, assessment and treatment is to develop risk monitoring plans and standards, track the implementation of risk management plan, use effective risk monitoring and risk response methods and means, report risk states, send out early warning signals of risks and propose risk treatment suggestions.

2.8.4 *Responsibilities in risk management*

2.8.4.1 *Responsibilities of the employer*

(1) Entrust the consulting agency to establish and implement suitable project risk management procedures, so as to supervise the project risk management work in the whole process of consultation.
(2) Review and approve the project risk management plan proposed by the consulting agency, including project risk assessment criteria.
(3) Participate in project risk assessment and follow up meetings.
(4) Actively promote relevant organizations to carry out project risk management, follow and properly deal with project risk management results in a timely manner.
(5) Organize the study on the risks (e.g. incomplete consultation documents, drawings and data submitted, and late submittal) to be treated by the employer in a timely manner according to project risk assessment results, provide and make decisions on suitable and feasible risk treatment measures, including risk control and mitigation measures and risk response plans, to mitigate the risks of the project.
(6) Have the final decision-making power on project risk assessment results and project risk management, and if needed, make the final decision by referring to the comments provided by the consulting agency.

(2) Responsibilities of the consulting agency
The responsibilities in risk management of railway engineering consultation may be further divided into responsibilities of the project manager and responsibilities of the project risk management director.

2.8.4.2 *Project manager*

(1) The project manager should conduct internal review and approval of the project risk management plan, including project risk assessment criteria, and formally submit the same to the employer.
(2) Participate in project risk assessment and follow up meetings.
(3) Promote the on-schedule execution of relevant project risk management work in accordance with the control management procedures and project progress established.
(4) Confirm the project risk treatment principals designated by the project risk management director according to project risk management results and reports of the project risk management director, follow up relevant project risk record forms provided by these project risk treatment principals, and get information about and follow up their project risk treatment progress and results through monthly reports on progress and monthly meetings. For risk items to be treated by the consulting agency, such as project progress plan and consultation review matters, provide and implement suitable and feasible risk treatment measures with the assistance from the project risk management director, including risk control and mitigation measures and risk response plans, to mitigate the risks of the project.
(5) Make adequate and suitable mutual coordination and arrangement for and give treatment to project risk management work with the employer and other relevant internal and external organizations in accordance with item (4) above.
(6) Review and approve project risk assessment results. If needed, make coordination with the project risk management director by referring to project risk record forms and provide professional consultation comments for the employer, so that the employer

can make final decisions on project risk assessment results and project risk management.

2.8.4.3 *Project risk management director*

(1) Prepare project risk management programs and establish and implement rational project risk management procedures to supervise the overall project risk management work, and propose the project risk assessment criteria.
(2) Execute the project risk management work on time in accordance with project progress and requirements, including presiding over project risk assessment and follow up meetings, and complete project risk record forms.
(3) Designate the project risk treatment principal for each risk in accordance with project risk assessment results, and confirm with the project manager. Update the project risk record forms periodically in accordance with the progress and result of project risk treatment, report the progress and result of project risk management to the project manager, and remind relevant organizations of following up the matters related to project risk management that need timely follow up.
(4) Monitor the risks and abnormalities in the project execution process, and if needed, provide professional consultation comments for the employer through the project manager and by referring to project risk record forms, so that the employer can make final decisions on project risk assessment results and project risk management.

2.8.5 *Risk management procedures*

Risk management procedures may be divided into risk analysis and risk management.

(1) Risk analysis
The purpose of risk analysis is to identify potential risks as early as possible, assess the risk levels, understand the potential risks of project activities and their influence, and get information about the risk bearing capacity of each party. Determine the priorities

of risk treatment based on risk levels and input suitable resources more effectively to control and mitigate risks.

The risks of decision-making, construction period, investment and technology are mainly considered in risk analysis.

(2) Risk management

Risk management may be accomplished by utilizing risk record forms. The purpose of risk management is to ensure that a uniform management method could be used to treat risks in all risk management stages.

The contents of risk management include clearly specifying the duties and responsibilities of relevant personnel, identifying and recording project risks, carrying out the follow up activities after identification of project risks, and effectively monitoring risks and eliminating risks as early as possible.

2.8.6 *Risk management methods*

(1) Communication management

Correct communication and coordination are decisive to successful implementation of a project. Communication consists of the communication between the consulting agency and other parties, such as the employer, the design institute and the construction contractor, and the internal communication of the consulting agency. The internal communication of the consulting agency includes personnel communication, information exchange and document delivery.

The project manager is the key to communication, also the bridge for effective internal and external communication. The project manager should adopt the correct communication mode by distinguishing changes of the external environment to avoid unfavorable influence of the external environment on consultation work. The project should be acutely aware of unfavorable atmosphere and mood in the construction process, control such atmosphere and mood with formal communication timely and correctly, and avoid the intensification and spreading of such atmosphere and mood to the maximum extent.

The consulting agency should (i) realize network information sharing by establishing a standard project information report system and information report format; (ii) strictly divide the sharing and confidential information, establish and adopt relevant systems to standardize such information; (iii) clearly specify the persons in charge of preparing and modifying the communication plan to ensure accurate and effective transfer of information and effective communication feedback mechanism; and (iv) clearly specify the method, time and format for work report from consultation personnel to the project manager, from the project manager to the employer and of relevant personnel.

(2) Comprehensive risk management

(i) Whole-process management

Manage uncertain factors of the project in its whole service life. Predict major risks that may affect the project objectives, find out the risks of and possible difficulties in objective achievement, and lay emphasis on in-advance identification, assessment and prevention measures. Break down the risk analysis, predict the possibility and rules of risk occurrence, and study the degree of effect or sensitivity of various risks on project objectives. Study the preventive measures for such risks as risk reserve and alternative technical proposal. Establish risk control measures such as risk monitoring system, risk identification, risk prediction and risk management assessment.

(ii) All-round management

Analyze the effect of risks on all aspects of the project, including construction period, cost, process, contract, technology and plan. Use comprehensive means to determine solutions in contractual, economic, organizational, technical and managerial aspects.

(iii) Organizational measures

Send specially-assigned persons to manage the risks having significant influence and give them corresponding responsibilities, authority and resources. Comprehensively implement the risk control accountability system, establish the risk control system, consider risk management as one of the tasks of consultation personnel at all levels of the consulting agency, improve the risk awareness of all members and strengthen risk monitoring.

(3) Talents and trainings
Select qualified persons and select the technicians with rich experience in survey, design and consultation as the chief consultant and discipline consultation directors. Provide technical support for consultation work by setting up an expert database, etc. Strengthen technical and managerial trainings in the consultation process.

(4) Complete plans
Make the progress plan, cost plan, allocation plan for manpower and material resources, risk plan and quality plan for the project, and systematically arrange all consultation activities according to these plans.

Determine the scope of all tasks necessary for achieving project objectives, assign the responsibilities, prepare the time schedule for each task, and clearly specify the manpower, material and financial resources necessary for each task. Determine the work specifications and standards for the project, which will be used as the basis and guidance for project implementation. Clearly specify the scope of responsibilities and the objectives, methods, approaches and time limits of work for the consultation personnel and their corresponding authorities. Guarantee the exchange, communication and coordination between the consultation personnel and the employer and design institute to get consistent opinions on every piece of work, increase the employer's satisfaction and provide conditions for follow up control of the project. Direct the consultation work in accordance with the approved consultation plan.

Chapter 3

Methods for Railway Engineering Consultation

3.1 Overview

Railway engineering has a large scale and wide influence, involving multiple disciplines, domains, technical policies, development planning, technical standards, technical principles, design concepts, key technologies and implementation schemes. During railway engineering consultation, the consulting agency needs to provide independent, comprehensive, complete and accurate advisory opinions on safety, reliability, applicability, cost-effectiveness, durability, systematicness and interface relations of railway engineering, applying scientific and systematic methods to guarantee that the safety, quality, construction period and investment of railway engineering are always under control and to fulfill the requirements of overall construction objectives.

Based on scientific principles of epistemology and methodology, as well as the characteristics of and actual experience in railway engineering consultation, the methods for railway engineering consultation mainly include the investigation method, observation method, experiment method, simulation method, question-raising method, induction method, analogy method, deduction method, systematic method and feedback verification method.

3.2 Stage Division of Consultation Work

According to the work steps and scope, railway engineering consultation is divided into three stages, including the consultation preparation stage, the consultation implementation stage and the consultation result forming stage.

3.2.1 *Consultation preparation stage*

The consultation preparation stage is the starting point of the consultation work and also the important link to carry out technical reserve for railway projects under consultation. The investigation method is the main method applied in this stage.

The investigation method is applied to widely and completely collect the technical data of similar projects, the preliminary planning and schemes of projects under consultation, the existing engineering technologies, especially the successful experiences as well as lessons drawn from failure of existing projects. Relevant information is collected and mastered in a comprehensive way through investigation to create good conditions for implementing the consultation work at a high starting point.

3.2.2 *Consultation implementation stage*

The consultation implementation stage is a work process for carrying out review, analysis and supplementary investigation for design documents and projects under consultation. The investigation method, observation method, experiment method, simulation method, question-raising method and systematic method are mainly applied in this stage.

The investigation method is applied to carry out supplementary and in-depth investigation based on the consultation needs and problems identified. The observation method is applied to identify the differences between actual project conditions and survey and design as well as relevant problems to obtain perceptual knowledge through reviewing design drawings and documents as well as verifying actual project conditions at project sites. For major projects, special structures and key technologies under consultation, the experiment method is applied to carry out tests and inspections and the simulation method is applied to carry out model tests and computer numerical simulation analysis to deepen the consultation work. The question-raising method is applied to standardize consultation operations and carry out complete and systematic sorting, verification and review for design documents and projects under consultation. The systematic method is applied to check safety, reliability, applicability, overall technologies, technical schemes, disciplines and engineering interfaces of projects, and to optimize engineering design and schemes in a systematic way.

3.2.3 *Consultation result forming stage*

The consultation result forming stage is the last link of the consultation work, and is a process of collecting and sorting out main conclusions and results obtained from the reviews and analyses at the consultation implementation stage as well as supplementary investigations. The induction method, analogy method and deduction method are the main methods applied in this stage. In order to guarantee the quality of the consultation work, it is necessary to verify, analyze, comprehend and improve the consultation results again by mainly applying the feedback verification method as well as the experiment method and simulation method if necessary.

The induction method is applied to conclude and summarize the problems identified, to predict and analyze the possible adverse effects, to identify the possible universal problems of projects, and to make suggestions on optimization, improvement and modification. Based on a full understanding of projects, the analogy method and deduction method are adopted to apply successfully engineering technologies and experience as well as lessons drawn from failure to consultation. The feedback verification method is adopted to solicit opinions from the employer and design institute, to re-evaluate the influences of consultation opinions against projects, to identify non-conforming consultation results, and to inspect the rationality and stringency of consultation results through logical analysis. Major technical schemes and measures proposed in consultation may be verified through tests and simulations if necessary.

3.3 Investigation Method

3.3.1 *Definition*

The investigation method means that a consulting agency understands the conditions of railway projects under consultation and similar projects in a planned and purposive way and collects relevant materials and data through direct and indirect forms including inquiry, network inquiry, meetings, data collection, books and documents, so that the agency is able to use existing experience and technologies on railway engineering design, research and construction as well as technical policies, development directions, technical standards, technical principles, design concepts, key technologies, implementation schemes and operational situations of similar projects for reference, to know the preliminary planning, schemes and approval documents of projects under consultation, to avoid common deviations, errors, omissions and conflicts in design, and to prevent project quality accidents.

Investigation is the basic work for consultation. Only a systematic grasp of the preliminary conditions of projects under consultation can guarantee a correct understanding of technical policies and standards. Only a systematic collection of technical data of similar railway projects, a close tracking of technology trends, a deep understanding of technical background and an overall grasp of technical methods can make up for each other's deficiency and lay a solid foundation for the smooth implementation of consultation. Only a comprehensive and deep understanding of quality and safety problems encountered in design and engineering of similar railway projects can prevent same or similar problems from happening again. Namely, only a scientific, systematic, comprehensive and serious investigation can guarantee comprehensive, systematic and reasonable understanding and analysis on project safety, reliability, integrity, durability, adaptability, maintenance, detail treatment, construction schemes and relevant project influences, and improve consultation quality.

3.3.2 *Classification*

According to the purposes of investigations, the investigation method is classified into descriptive investigation method and explanatory investigation method. According to time and project environmental types, the investigation method is classified into longitudinal investigation method and lateral investigation method. According to content characteristics, the investigation method is classified into overall investigation method and interrelation investigation method.

3.3.2.1 *Descriptive investigation method*

The descriptive investigation method is mainly about "what is it." New problems are identified through investigations, and the key points and major problems of the engineering consultation to be implemented are summarized through collecting, sorting and compiling the preliminary planning, scheme research and project approval documents of projects under consultation as well as documents, experience introduction, technical principles, design concepts, key technologies, implementation schemes and operating conditions of similar projects. For example:

Before implementing consultation for bridge deck arrangement of ballastless tracks, it is known through investigations that the space between bridge deck ballastless track and protection wall and the space between two ballastless tracks may or may not be filled and leveled with ballast. It is also known that these spaces are normally filled and leveled with ballast for high-speed railways in Germany, while they are not filled and leveled with ballast for Shinkansen in Japan as shown in Fig. 3.1. However, it is unable to answer why these spaces are or are not filled and leveled with ballast during descriptive investigations.

3.3.2.2 *Explanatory investigation method*

The explanatory investigation method is mainly about "why", and reasons are explained through investigations. For problems identified in data and compilation of investigation results, consultants

(a) Both sides of ballastless tracks of high-speed railways in Germany are filled and leveled with ballast.

(b) Both sides of ballastless tracks of Shinkansen in Japan are not filled and leveled with ballast.

Fig. 3.1 Two Bridge Deck Forms of Ballastless Tracks Abroad

carry out discussions, targeted data inquiry and other activities to understand their causes. Investigations are not only for knowing problems intuitively, but also more importantly for discussing the causes of problems. Therefore, the beginning of explanatory investigations is normally the result of descriptive investigations. For example:

Investigations are carried out to determine whether it is necessary to fill and level the bridge deck of ballastless tracks with ballast. It is known from further investigations that Germans think if the space between bridge deck ballastless track and protection wall and the space between two ballastless tracks are filled and leveled with ballast, the ballast will be able to absorb part of the kinetic energy after train derailment and reduce train speed, so it is beneficial to prevent train overturn and reduce casualty loss. The bridge deck filled and leveled with ballast may also be used as a rescue passageway. In case of evacuation and rescue during an emergency stop, the filled and leveled bridge deck is able to avoid the inconvenience caused by the height of steps, so that passengers on train may be evacuated quickly, and rescue personnel may access and exit quickly. It can save a lot of time, especially during the evacuation of the old, weak, sick, disabled and young. Filling and leveling with ballast is also able to protect supporting layers and reduce maintenance load, and it is beneficial to maintain the transverse stability of supporting layers and track bed slabs. Japanese railways take the road of light-weight ballastless track. The bridge deck load may be reduced without ballast filling and leveling, and derail accidents are prevented mainly by improving performance of trains.

In the meantime, it is identified through investigations that ballast will increase the secondary dead load which increases work amount and cost of girder body, and ballast itself requires more investment, and measures must be taken to prevent ballast from splashing when trains are running at a high speed. Ballastless tracks' requirements on bridge deck drainage are restricted, so the influences against bridge deck drainage by ballast also need to be deeply analyzed. In addition, there is no actual case that is able to verify the actual effects of ballast filling and leveling in preventing train overturn and reducing casualty loss, and it is difficult to conduct simulation with laboratory facilities.

3.3.2.3 *Longitudinal investigation method*

The longitudinal investigation method refers to the investigation and comparison for characteristics of a certain project within a long period of time. For example:

Through investigations for the development of ballastless tracks in Germany, it is known that Deutsche Bahn AG, university institutes and industrial communities in Germany have been promoting the research, development and application of ballastless tracks since the 1970s, and ballastless tracks in various structures and types have been applied in German railways. There are over 20 enterprises participating in the development of new structures of ballastless tracks in Germany, which forms a competitive market situation and pushes forward new technology development, and the structural forms proposed are of a great diversity. It is specified in Germany that before being approved by the Federal Railroad Administration and then popularized and applied, new ballastless track structures must pass a five-year operation trial on a line with a speed of 230 km/h and above after passing tests. The frequently used ballastless track structures in high-speed railways in Germany include Bögl type, Rheda type, Züblin type, Atd type, Getrac type and Berlin type. Rheda type ballastless tracks are most widely applied, and Züblin type and Rheda type tracks are similar in structure but different in construction methods.

In 1972, the former Federal Republic of Germany laid a kind of sleeper-type ballastless track at Rheda Station, and since then this series of ballastless tracks (Rheda System Solid Track) were named after Rheda in Germany. Rheda type tracks evolved from the traditional Rheda type to Rheda-Berlin HST V1 type, Rheda-Berlin HST V2 type, Rheda-Berlin HST V3 type and Rheda 2000 type, and the structural height keeps reducing as shown in Fig. 3.2. Based on many operation trials in trail laying sections as well as long-term observations and researches, Rheda type ballastless tracks have been widely popularized and applied on high-speed bridges, tunnels and soil subgrade in Germany.

3.3.2.4 *Lateral investigation method*

The lateral investigation method refers to the investigation for characteristics of a certain type of railway projects within a specified time period. For example:

Through investigations for how to determine the inner contour line for tunnel lining and select its dimension in the engineering design of high-speed railway tunnels in different countries, it is known that the inner contour line dimension for tunnel lining of high-speed railways in China is determined according to tunnel construction clearance, number of tracks, distance between centers of tracks, tunnel equipment space, aerodynamics effects, structural forms of tracks as well as operating and maintenance modes. For high-speed railways with a design running speed of 300–350 km, the effective clearance area of double-track tunnels shall not be less than $100\,\mathrm{m}^2$, and that of single-track tunnels shall not be less than $70\,\mathrm{m}^2$ (see Fig. 3.3).

The clearance area in tunnels of high-speed railways in Germany is determined according to the requirements of aerodynamics effects (remission and consumption) on effective clearance area or blockage

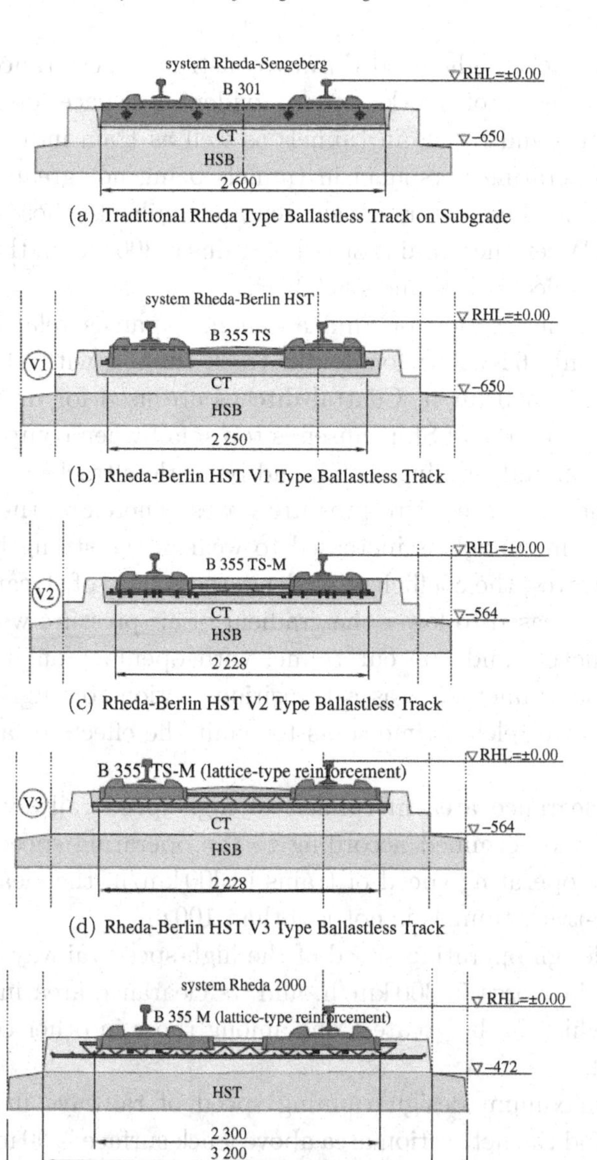

(a) Traditional Rheda Type Ballastless Track on Subgrade

(b) Rheda-Berlin HST V1 Type Ballastless Track

(c) Rheda-Berlin HST V2 Type Ballastless Track

(d) Rheda-Berlin HST V3 Type Ballastless Track

(e) Rheda 2000 Type Ballastless Track

Fig. 3.2 Rheda Type Ballastless Track of Germany (Unit: mm)

ratio of tunnels, high-speed railway construction clearance, distance between centers of tracks, curve widening, space dedicated for maintenance and rescue in tunnels as well as train induced air flow born by operating personnel in tunnels being not greater than the critical value. The factors considered are similar to those considered in China. When the running speed of trains is 300 km/h, the clearance area in double-track tunnels is 92 m^2.

The tunnel section of Shinkansen in Japan is relatively small, and it is only 62–64 m^2 for double-track tunnels with an operating speed of 240–300 km/h. Central ditches are used for drainage. The tunnel section area of Shinkansen is too small when compared to the train section and running speed, and ballastless tracks are unable to absorb and dissipate micro-pressure waves. Therefore, the roughness of sections in tunnels is increased to weaken the strength of micro-pressure waves, the coefficient of slenderness ratio of streamline train heads is increased to lower the gradient of air pressure wave in front of train heads, and one cut tunnel with openings on its side wall is added at tunnel exits as a transition section to lengthen the air pressure wave release time so as to remit the effects of air pressure waves.

The clearance area in tunnels of high-speed railway in France is normally determined according to the operating speed of trains. When the operating speed of trains is 300 km/h, the clearance area in double-track tunnels is not less than 100 m^2.

The design operating speed of the high-speed railway from Seoul to Busan in Korea is 300 km/h, and its clearance area in tunnels is 107 m^2, which is the greatest one among those in other countries in the world.

The maximum design running speed of railways in Taiwan is 350 km, and the net section area above track surface is 90 m^2. In order to prevent explosive noises caused by trains when exiting tunnels, 45° skew tunnel portals are adopted for tunnels with a length over 3,000 m. Open cut tunnels for depressurizing are set, and openings are set at top of these tunnels to gradually release micro-pressure waves.

3.3.2.5 *Overall investigation method*

The overall investigation method refers to investigations that aim at understanding overall characteristics of projects, emphasize the representativeness of the object investigated to acquire relatively reliable overall characteristic parameters. For example:

Through investigations for the overall characteristics of high-speed railway projects in China, it is known that the technical standards for a high-speed railway in China are determined through a comprehensive comparison and selection guided by a system optimization principle based on the railway's functions in railway networks, topographical and geological conditions along the line, carrying capacity and transportation demands. The design speed is determined through a comprehensive technical and economic comparison based on a project's functions in rapid passenger transportation networks, transportation demands and engineering conditions, and it shall meet the requirements of traveling time target value. A double-track bi-directional electrified railway is adopted. The distance between main tracks, the minimum plane curve radius and the maximum gradient are determined according to the design running speed, traffic organization modes as well as requirements on safety and comfort. The length of the receiving-departure track is 650 m. The type of high-speed trains or EMUs fits the speed of passenger trains. The train operation control modes of high-speed railways include CTCS Class 2 train operation control system (standby mode) based on track circuit transmission as well as CTCS Class 3 train operation control system based on GSM-R radio communication transmission. A centralized traffic control system is adopted as the train traffic control mode. The minimum headway is determined to be three minutes according to transportation demands.

For high-speed railways in China, the emphases are laid on standard matching and coordination among systems including civil engineering, traction feeding, train operation control, high-speed train, operation dispatching and passenger service; on the interface coordination; on matching and compatibility of fixed and mobile facilities; on system optimization; on coordination between civil

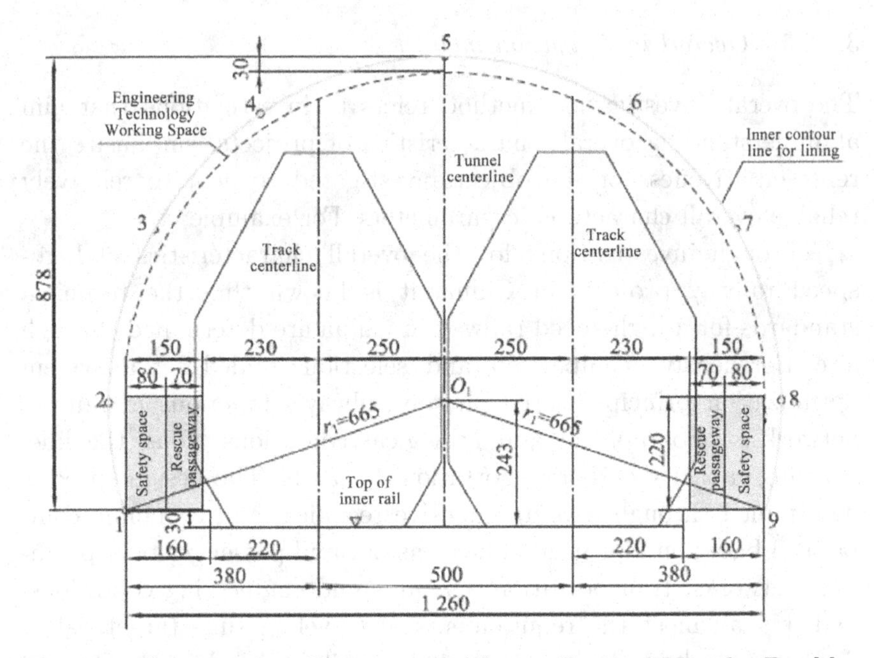

Fig. 3.3 Construction Clearance and Inner Contour Line for Double-track Tunnels of High-speed Railways with a Speed of 350 km/h in China (Unit: cm)

works; on coordination between civil works and associated works; on interface coordination between the works of a specific line and the line's associated works as well as between the works of the line and the line's adjacent railways; on comprehensive optimization of quality, safety, construction period, investment, environmental protection and scientific innovation; on establishing a precise measuring network consisting of survey and design network, construction network and operation and maintenance network; on strict control of settlement of main structures including subgrade, bridges, tunnels and buildings; on adopting applicable engineering measures to control noises caused by wheel/rail system, pantograph and overhead contact system, electro-mechanical system and aerodynamics.

3.3.2.6 *Interrelation investigation method*

The interrelation investigation method refers to investigations that aim at analyzing the interrelation of some projects or project conditions to acquire reliable judgment on the interrelation. For example:

Before implementing consultation for railway projects in loess areas, through the investigations for interrelation among railway projects, loess and loess's natural factor influences, it is known that loess in collapsible loess areas is of a great porosity. If the weather is dry, the moisture content in soil in low, overhead pores will develop, soil structure will become loose, and the compressibility of soil will be high. Meanwhile, the compressibility will change along with the change of moisture content in soil. The collapsibility is the main problem for railway projects in loess areas, and if measures for handling subgrade basement subsidence are not appropriate, the safety of railways will be seriously affected. Reinforcement measures for subsoil and side slopes are determined according to the cause of formation, characteristics, distribution and thickness of loess. The treatment for collapsible soil layers at basement of culverts, measures for waterproofing between culvert segments as well as seepage-proofing works for paving of culvert beds at inlet and outlet sections of culverts will greatly affect the quality of railway projects.

Underground cavern systems with complicated forms and runoffs are usually formed in loess due to centralized infiltration and subsurface erosion caused by surface water, and they pose threats to the stability and safety of railway subgrade, side slope, tunnel body and other works. Loess cratering (see Fig. 3.4) is of complicated forms, concealed and hard to locate, and it also develops fast. Due to human factors or centralized drainage and seepage of surface water, new cratering may soon form (even during operation) or existing cratering may expand rapidly. Therefore, cratering is a kind of unfavorable geological phenomenon with dynamic development and a major geological problem affecting safety of railway projects.

For a landform dominated by loess ridges, loess is distributed at surface of ridges in a draping shape, and bedrock is below loess. The loess-rock interface leans to river valleys and is usually located at free faces, consisting of the basic geological conditions for highly developed loess landslides (see Fig. 3.5). Once existing landslides are

Fig. 3.4 Loess Cratering

Fig. 3.5 Loess Landslide

triggered or new landslides are generated due to improper setting of railway projects, the safety of railways will face serious threats.

In case of steep hillsides with a large number of landslides and slope slide developed at higher locations, and waste slag is stacked along ravines, once there is abnormal rainfall, it will be easy to cause silting in ravines or even large-scale mudflow and debris flow rushing out of valleys, especially narrow V-shaped ravines such as mud ravines and back gullies. Once ravines are blocked, barrier lakes may form, which will pose serious threats to the safety of railways if clearance below railway bridges is not enough.

For tunnels passing through loess formations, there will be relatively serious potential safety hazards if their arch top is located in the boundary of loess and rock.

There are many factors affecting the engineering properties and physico-mechanical indicators of loess, and the main factor is the age. The engineering properties and physico-mechanical indicators of loess formed in different ages and having different causes of formation are of obvious differences, which in different degrees affect the accuracy of engineering designs and the practicability of engineering measures. It is necessary to determine targeted schemes for various and complicated characteristics of loess to make engineering measures become abundant.

Earthquakes may cause stability loss of unstable loess slopes, trigger existing landslides or generate new landslides, and subsequently form surface fracture zones, saturated loess liquidation and other geological hazards. Therefore, it is necessary to pay special attention to the possible safety hazards that may be induced by loess slopes and existing landslides within a certain range at both sides of railways under seismic actions.

Water may induce geological hazards in loess areas, so good drainage facilities may guarantee the stability of subgrade side slopes and subgrade. The subgrade drainage shall be joined with the drainage equipment of neighboring bridges, culverts and stations to constitute a reasonable drainage system. In the meantime, comprehensive agricultural water-conservation application shall be taken into account to prevent the farmland from absence of irrigation and destruction.

Scour prevention measures, anti-seepage measures and other reinforcement measures shall be taken for drainage ditches in loess section based on the poor water-resistance performance and collapsibility of loess.

3.3.3 *Characteristics and precautions*

(1) The investigation method is different from the observation method and experiment method. It acquires the indirect experience of past and other projects through other channels in an indirect way, instead of acquiring the conditions, parameters and characteristics of projects under consultation. Thus, the analysis and identification of consultation personnel as well as practical inspections are necessary. It is not possible to practice everything, so the investigation method is indispensable in consultation.

(2) Objectivity is the basic principle in investigations and the essential condition to guarantee successful investigations. Human interventions are not allowed during investigations, and the status, parameters and relevant characteristics of investigation objects must be recorded objectively.

(3) The investigation method is able to collect abundant data at the same time, and it is easy to use and of a high efficiency. It is able to obtain clues from preliminary schemes of projects or from existing projects to solve problems and broaden consultation ideas. Circumstantial evidence may also be obtained through investigations to become references for consultation and help to solve similar problems. Engineering experience requires a long-term accumulation, and the investigation method is an effective way for identifying and accumulating data as well as improving technical level of consultation.

(4) A comprehensive point of view must be established during investigations, because only comprehensive investigations are able to guarantee a correct understanding of engineering technologies. The relationship between individual and general as well as between special and universal shall be properly handled. General rules shall be uncovered by starting with typical problems. Multiple methods shall be adopted when classifying and processing investigation material to reflect essences and patterns as complete and accurate as possible.

(5) There must be specific purposes and plans for investigation to get twice the result with half the effort as well as complete and accurate data. Investigation personnel shall give full play to their subjective initiatives, think independently, research and identify carefully, and distinguish truth from falsehood during investigations.
(6) Although obtaining investigation material and using the material for implementing consultation work are two different links, it is always necessary to combine them during actual work, i.e. using investigation material to implement consultation based on certain material obtained, carrying out investigations during consultation as necessary, and guiding re-investigations with consultation demands.

3.4 Observation Method

3.4.1 *Definition*

The observation method means consultation personnel designedly carry out project site verification and review design drawings and documents by using senses, instruments and equipment to obtain perceptual knowledge based on the railway engineering technical standards, theories, experience and approaches they grasped, the overall construction objectives and requirements on safety, quality, construction period and investment of projects under consultation as well as the responsibilities agreed in contracts.

Observation is a process to form perceptual knowledge through the senses, and then a process by which the brain organizes or connects the knowledge using certain ways. That is to say, the observation is a combination of perceptual knowledge and certain organization forms. Observation includes not only checking project sites and design drawings documents, but also understanding them rationally. It is not only accepting external information, but also preliminarily processing the information.

3.4.2 *Classification*

According to the means and methods of observation, the observation method is classified into the direct and indirect observation methods; according to the properties and contents of observation, the observation method is classified into the qualitative and quantitative observation methods.

(1) The direct observation method refers to sensuous observation, i.e. directly observing, checking and recording project site conditions, appearance, design drawings and documents (see Fig. 3.6 for an example of site verification by consultation personnel). It is the most primary method in railway engineering consultation. For example:

  (i) During the preliminary stage and consultation process of railway engineering design consultation, topography, landform, roads, rivers, ditches and pipelines at project sites as well as major works, large temporary works and large reconstruction works critical to cross and vertical sections of lines are observed, recorded and verified. Where:

   - During the site verification for bridges, verifying conditions of roads, existing railways and pipelines that railway bridges cross as well as overview of rivers that railway bridges cross, and checking geological conditions and bridge site conditions;
   - During the site verification for tunnels, verifying the unfavorable geology and drainage at tunnel portals; water conservation and environmental protection conditions; construction site conditions at tunnel portals as well as construction road conditions of main tunnels and inclined shafts;
   - During the site verification for girder fabrication yards, verifying the position, landform, area and girder transport road of yards;

Fig. 3.6 Site Verification by Consultation Personnel

- During the site verification of track laying base, verifying the base scale, conditions for connecting with existing stations and interlocking relations of turnouts at connection points;
- During the site verification for stations, verifying station position layout conditions and engineering scale.

(ii) For another example, observing, recording and verifying design drawings and documents of each discipline during consultations for engineering design of high-speed railways. Where:

- For the route discipline, observing, recording and verifying route plans and profile diagrams (and whether they matched); safety protection measures, construction methods, construction transition schemes and construction procedures for crossing buildings.
- For the geology discipline, observing, recording and verifying the drill hole sampling quantity; various *in situ* tests, test items and test results; test pile results (and whether they are used in shop drawings); physical

Fig. 3.7 Unfavorable Geology–Karst Cave

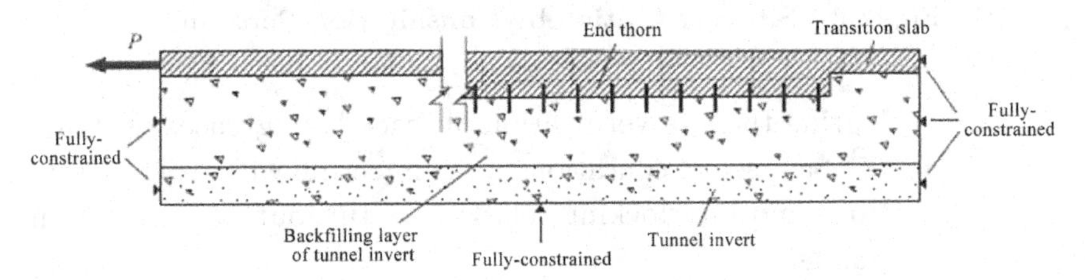

Fig. 3.8 Scheme for End Thorns of Plate-type Ballastless Tracks of Long-span Bridges at Bridge–tunnel Connection Sections

mechanics parameters of rock and soil; engineering geological changes caused by project construction as well as prediction of geological problems. Figure 3.7 shows an example of unfavorable geology.

- For the track discipline, observing, recording and verifying laying sections and structural form selections of ballastless tracks; the design of seamless tracks on special bridges and turnout areas. A scheme for end thorns of plate-type ballastless tracks is shown in Fig. 3.8; structural design, laying technology standards, and design of vibration and noise reduction and insulation measures for high-speed seamless turnouts on viaducts; interfaces between track works and associated disciplines.

Fig. 3.9 Cutting with High Side Slopes

- For the subgrade discipline, observing, recording and verifying settlement control standards and measures after subgrade works; reinforcement measures for soft soil and mollisol subgrade; the construction period and construction organization of subgrade preloading; reinforcement and protection measures for deep cutting side slopes (see Fig. 3.9), dangerous rocks and rockfall; the design of transition sections among subgrade, bridges, tunnels, culverts, ballast track and ballastless tracks.
- For the bridge discipline, observing, recording and verifying the rationality of span and abutment arrangement of bridges; the longitudinal and transverse stiffness of bridge piers and abutments. A steel-concrete combined rigid frame pier is shown in Fig. 3.10; the stiffness of each abutment and foundation of a certain bridge; the vehicle–bridge interaction analysis and conclusion of special bridges; the seismic calculation and seismic measures of bridge structures; the construction scheme design of bridges; the hazard prevention, rescue and maintenance design of bridges.

Fig. 3.10 Steel–concrete Combined Rigid Frame Pier

- For the tunnel discipline, observing, recording and verifying waterproof and drainage measures; handling principles and engineering measures for karstic water in tunnels with karst developed; construction schemes for weak surrounding rock of tunnels; the engineering design for advance geological forecasts and safety emergency plans; engineering measures and setting standards for remitting aerodynamics effects of tunnels; setting principles of fire rescue systems. An unfavorable topography at tunnel portal-eccentrically compressed tunnel portal is shown in Fig. 3.11.
- For the station discipline, observing, recording and verifying the configuration of necessary safety equipment for station operation; the scale of main equipment; the station arrangement; the comprehensive design of various cable trenches and pipelines. Figure 3.12 shows ballastless tracks in station area of high-speed railways.
- For the electrification discipline, observing, recording and verifying external power interfaces; the design for

Fig. 3.11 Unfavorable Topography at Tunnel Portal-Eccentrically Compressed Tunnel Portal

reservation of support foundation and backstay foundation of OCSs on subgrade and bridges; design schemes for drop tubes of OCSs in tunnels, for introducing OCSs into the urban area of terminals within consultation scope and for grounding and lightning protection; suspension schemes for OCSs in EMU operation maintenance depots, EMU depots, high-speed storage yards and EMU service shops. Figure 3.13 is a picture of the autotransformer.

- For the EMU and rolling-stock equipment discipline, observing, recording and verifying the setting of EMU operation and maintenance facilities as well as locomotive and rolling stock facilities for EMUs; names and functions of tracks configured for EMU operation and maintenance facilities as well as locomotive and rolling stock facilities; design and technical requirements on main

Fig. 3.12 Ballastless Track in Station Area of High-speed Railway

buildings of EMU servicing depots; EMU servicing and maintenance information management systems of EMU servicing depots. Figure 3.14 shows a three-dimensional diagram of EMU bogie replacement device.

- For the comprehensive maintenance discipline, observing, recording and verifying the distribution and scale of comprehensive maintenance organizations; general drawings of comprehensive maintenance facilities, layout plans of main workshops and equipment, and relevant technical requirements; the configuration and type selection of comprehensive maintenance equipment. See Fig. 3.15 for an example of a comprehensive maintenance car.
- For the electric discipline, observing, recording and verifying the type selection of equipment and material; the layout of electric pipelines; capacitive current compensation schemes for the entire cable power transmission

Fig. 3.13 Autotransformer

Fig. 3.14 3D Diagram of EMU Bogie Replacement Device

line along the track; power supply plans for snow-melting devices of turnouts on bridges and for stations; the type selection and installation of electric facilities in tunnels; the control of emergency lighting systems and fire doors as well as start of fire equipment; the interface matching for

Fig. 3.15 Comprehensive Maintenance Car

Fig. 3.16 Station Lighting Design

electric equipment and comprehensive grounding systems; lightning protection measures; technical parameters of lightning protection and grounding. Figure 3.16 shows a station lighting design.

- For the wire communication discipline, observing, recording and verifying channel requirements of comprehensive video monitoring systems on SDH transmission systems and IP data networks, and interface relations among them; site installation conditions of outdoor cameras at road overpass railways and key station areas; schemes for realizing linkage among SCADA, power sources, environmental monitoring systems, machine room lighting systems and other systems. Figure 3.17 shows a comprehensive video monitoring of station.

Fig. 3.17 Comprehensive Video Monitoring of Station

- For the wireless communication discipline, observing, recording and verifying constitution modes of GSM-R system networks; the setting of GSM-R base stations and relay equipment. Figure 3.18 shows an example of simulation analysis for communication network planning.
- For the electromagnetic compatibility discipline, observing, recording and verifying electromagnetic interference conditions and protection measures of various kinds of navigation equipment at stations and yards; protection schemes for wireless stations along lines; protection measures for oil and gas pipelines along lines as well as for construction safety.
- For the signal discipline, observing, recording and verifying the general structure of signal systems; the compatibility between the train control system and existing 200 km/h speed-increasing sections; the interface design for the signal system of a certain line and the signal system of its adjacent and associated railways and hubs; the performance of equipment of signal systems against electromagnetic interference, lightning interference and

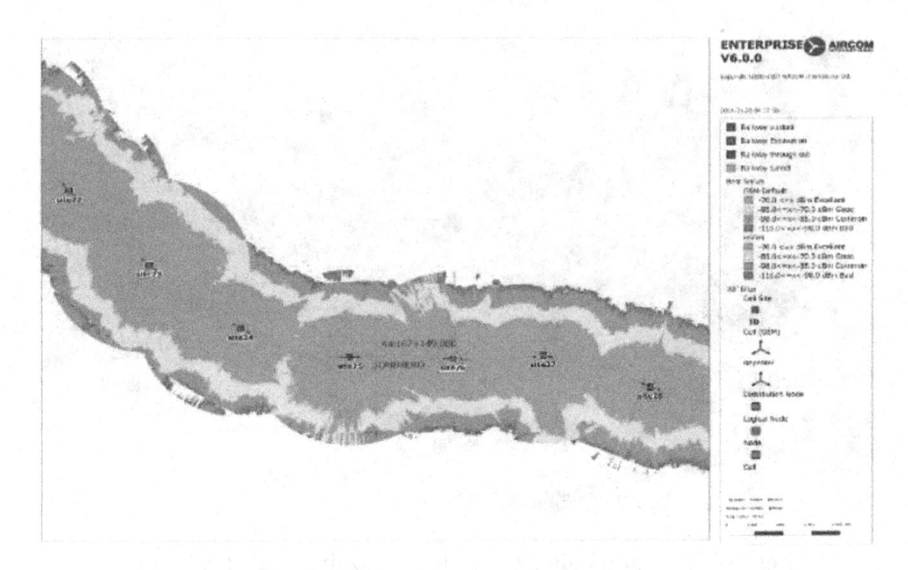

Fig. 3.18 Simulation Analysis for Communication Network Planning

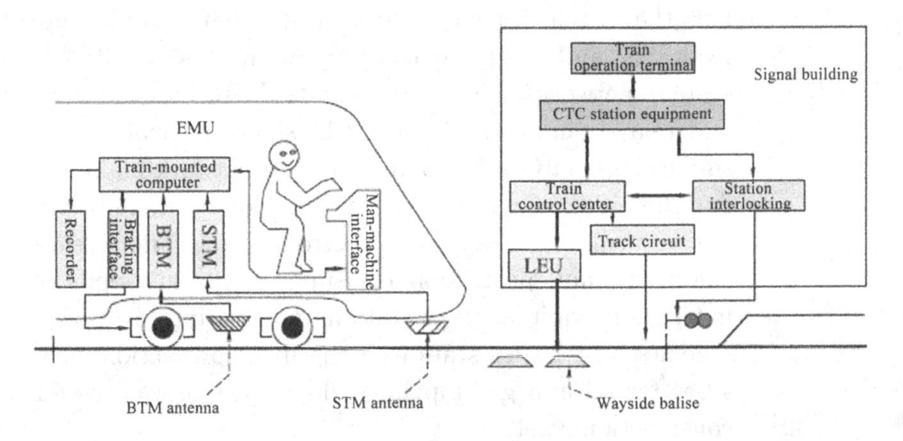

Fig. 3.19 Structure of CTCS Class 2 Train Control System

traction current interference; the safety and reliability of equipment of signal systems. Figure 3.19 shows the structure of CTCS Class 2 Train Control System.

- For the informatization discipline, recording and verifying automatic fare collection systems and passenger service systems;

Fig. 3.20 Clearance-invasion Monitoring Net

- For the hazard prevention and monitoring discipline, observing, recording and verifying automatic fire alarm systems in large spaces of stations; the setting of natural hazard monitoring points; power supply and communication modes of site equipment of foreign clearance-invasion monitoring systems (see Fig. 3.20) as well as hazard prevention and safety monitoring systems; installation positions of aerovanes; the setting of access control points; the lightning protection design for signal buildings; communication stations, substations, section intelligent box-type substations, large and super major station buildings as well as lamp bridges.
- For the building discipline, observing, recording and verifying the architectural image and decoration design of station buildings as well as the harmony with surrounding environment; the linking of station buildings and station area facilities; the passenger flow and directing systems; the design for green channels, landscapes and barrier-free facilities (see Fig. 3.21).

Fig. 3.21 Traffic Facilities of Station and Station Square

- For the structure discipline, observing, recording and verifying the engineering protection for deep foundation pits of large-scale station buildings and underground stations; main structures of large-scale passenger stations; curtain walls and decorative pendants; the waterproofing and drainage for roof; anti-skid floors; the steel structure painting; the working procedure exchange; the protection for aloft work and three dimensional operation; the safety protection for construction, operation and passengers at existing passenger stations.
- For the heating, ventilation and air conditioning (HVAC) discipline, observing, recording and verifying the heating, ventilation and air conditioning as well as measures for indoor water supply and drainage, for fire protection using water and gas, and for fireproof sealing.
- For the environmental engineering discipline, observing, recording and verifying engineering measures for ecological environmental protection as well as water and soil conservation of the entire line; the land rehabilitation;

Fig. 3.22 Water Supply Equipment for Passenger Trains

protective measures for noise and vibration sensitive spots along the line; prevention measures for electromagnetic pollution, solid waste pollution, atmospheric pollution and water pollution.

- For the water supply and drainage discipline (Fig. 3.22), observing, recording and verifying engineering measures for drainage pipes to cross station tracks, bridges and tunnels; the sewage treatment process and discharge; the construction organization of drainage works; transitional measures during construction.
- For the construction organization discipline, observing, recording and verifying construction schemes of full-span box girders (Fig. 3.23); transport and erection schemes of full-span box girders in bridge–tunnel connection sections; construction schemes of large-span bridges with special structures and these bridges' influences against the overall construction organization; construction schemes of long or extra-long tunnels and these tunnels' influences against the overall construction organization; the feasibility and

Fig. 3.23 Prefabrication Yard of Full Span Box Girders

reasonability of construction organization plans for track laying.

- For the engineering economy discipline, observing, recording and verifying quantities of land requisition and relocation works; key large-scale temporary works; material supply plans; transportation optimization schemes; land source conditions; bases of land requisition price; feasibility of transitional schemes for terminals. During the direct observation, the observation effects may be improved by adding observation angles, lengthening observation time and increasing observation times. The direct observation method is simple, direct, slightly restricted by objective conditions and not restricted by time. However, it also has limitations such as a limited scope of observation, a low accuracy and possible errors.

(2) The indirect observation method refers to the method in which consultation personnel use instruments and equipment to observe project site conditions, project appearance and design documents to enlarge the observation scope, improve the observation accuracy and complement the shortcomings of direct observation.

Fig. 3.24 Setting-out of Densification Reference Mark (GRP) for Type II Slab Track

The indirect observation method is one of the major means and approaches in railway engineering consultation. For example:

The settlement and deformation of railway projects are generally observed by construction contractors. Consulting agencies examine and verify observation schemes, implementation processes and measurement results of settlement and deformation, analyze, forecast and evaluate settlement and deformation, and use measuring instruments to carry out repetition measurement and parallel observation for a certain proportion of settlement and deformation. The repetition measurement is regularly or irregularly carried out for monitoring datum points and working datum points of settlement and deformation using precise digital level gauges and matching indium tile level rulers as per leveling lines of construction measurement as well as technical requirements of leveling measurement. The parallel observation is carried out according to the datum points and working datum points after the repetition measurement as well as the analysis, forecast and evaluation requirements on the settlement and deformation of railway projects under consultation, and by using precise digital level gauges and matching indium tile level rulers as well as measurement accuracy grades, measurement routes and deformation observation points being the same with those in construction measurement. The subgrade is the key monitoring point in the parallel observation, and sections that are under the monitoring of construction contractors and have relatively large differences in deformation are also observed according to the key engineering structure or sections proposed in analyses and evaluations. Figure 3.24 shows the setting-out of densification reference mark (GRP).

The indirect observation using instruments and equipment also has limitations, including wrong results caused by instrument and equipment issues, improper uses of instruments and equipment, or errors of instruments and equipment.

(3) The qualitative observation method refers to the observation for characteristics of project site conditions, project appearance, design drawings and documents as well as the qualitative relationship between those characteristics and relevant projects. The qualitative observation method is the most basic observation, and it is also the starting point and foundation of consultation work. For example:

It was observed through the site verification for a proposed high-speed railway bridge that the bridge is located at the departure side of a railway station and crosses the outer ring road of City A and the highway to City B, and the simply-supported box girders with a span of 32 m in the design is able to meet the requirements on crossing. When crossing the outer ring road of City A, there are high-voltage lines above the bridge, so the clearance between the high-voltage lines and the bridge was not able to meet the requirements on safety operation and erecting 32 m box girders with a bridge girder erection machine. Meanwhile, the high-voltage lines are adjacent to a substation, so the method of increasing the height of iron towers is not available. Therefore, it was concluded in the consultation that the high-voltage lines needed to be relocated. In addition, it was

(a) The high-speed railway bridge crosses the outer ring road of City A, with high-voltage lines over the bridge.

(b) The high-speed railway bridge crosses the highway to City B.

Fig. 3.25 Site Conditions of a Proposed High-speed Railway Bridge

identified that the clearance below the high-speed railway bridge was not enough when it crossed the highway to City B. Therefore, it was concluded in the consultation that the highway needs to be excavated or relocated. See Fig. 3.25.

(4) The quantitative observation method refers to the quantitative observation and description for engineering works or engineering design. It is based on profound qualitative knowledge and aims at grasping engineering characteristics accurately according to the relationship between quantities. For example:

In the above-mentioned indirect observation method, the repetition measurement and parallel observation carried out by

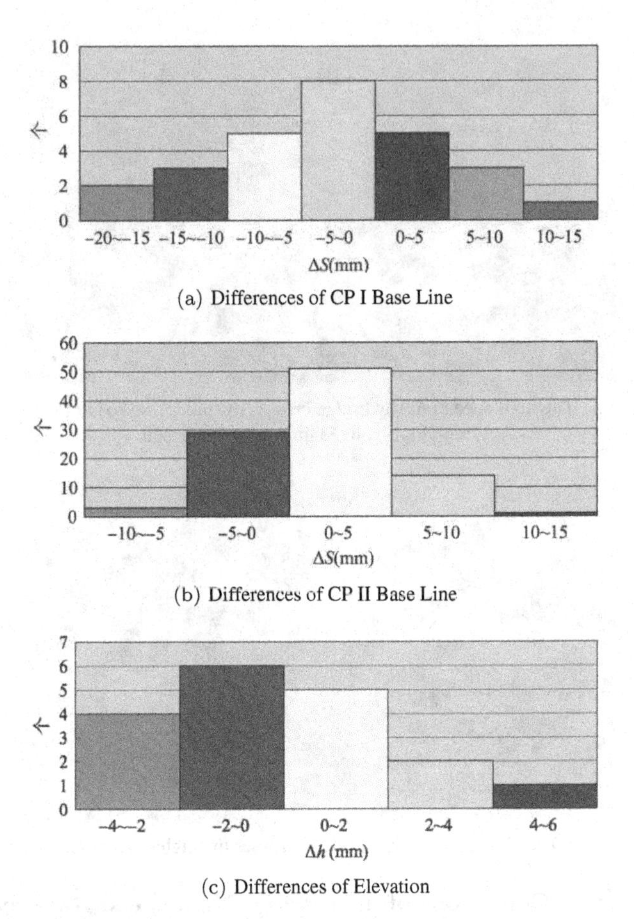

Fig. 3.26 Statistical Diagram for Differences in Parallel Observation by the Consulting Agency for a Railway Project

the consulting agency as per a certain proportion for observing the settlement and deformation by using measuring instruments belong to the quantitative observation method. See Fig. 3.26 for statistical diagram for differences in parallel observation by the consulting agency for a railway project.

3.4.3 *Characteristics and precautions*

(1) Observation is an important means for obtaining first-hand material. The direct observation is able to reveal actual problems

in projects. The indirect observation is able to expand the horizon, open the mind and deepen the understanding. The consultation quality may only be verified through observation. The most important difference between the observation method and the experiment method is that the observation method is free from human interference. Even if instruments are used, their basic forms and movement modes will not be interfered into or changed.

(2) The information needed may only be obtained by observing on purpose, and a detailed and practical consultation outline is necessary during site verification, design optimization, construction drawing approval, construction process consultation and special consultation. Things are observed or ignored according to the goals determined in the consultation outline to be concentrated. Necessary things may only be observed through elaborate plans. Comprehensive information as well as the relationship between the information and its associated things may only be understood through systematic observation.

(3) Observation does not interfere or change the basic forms and movement modes of things, so observation results are repeatable.

(4) The objectivity is the primary objective of observation, and that is determined by the objectivity of the physical world. It is necessary to observe and reflect according to the true features of the objects under observation, to proceed from reality in all work, to be practical and realistic and to prevent preconception.

(5) Observation is a complicated process of cognition, so its objectivity is always not accomplished in an action. During observation, it is necessary to grasp various relationships and forms of objective targets from different positions, angles and levels and to understand the targets comprehensively. Objective things are complicated, various and infinite, so it is necessary to select typical projects that are able to represent the common characteristics of a same type of projects and then determine other projects relating to the typical ones as the indirect observation targets.

(6) During observation, pay close attention to various details, make detailed observation records, determine observation scope and do not miss accidental events. Think actively and strengthen the relation with theories. Insist on the repeatability of observation. Reveal and then eliminate accidental errors in observation through observing repeatedly.

3.5 Experiment Method

3.5.1 *Definition*

The experiment method refers to consulting agencies using instruments and equipment to obtain engineering data under artificially controlled conditions or simulated engineering conditions for key technologies of design and projects as well as major issues concerning project safety, quality and investment.

3.5.2 *Classification*

Based on the relationship between quantity and quality, the experiment method is classified into the qualitative experiment method and the quantitative experiment method. Based on the functions of experiments during the process of cognition, the experiment method is classified into the factorial experiment method, the comparison experiment method and the intermediate experiment method.

(1) The qualitative experiment method refers to an experiment form that determines whether some factors affecting projects exist and whether some factors are related. For example:

The quality of pile foundation is directly related to the safety and durability of railway bridges. Pile foundation belongs to concealed works, so in order to guarantee the quality of railway projects, the pile body integrity is inspected qualitatively by using the low strain reflection wave method and the acoustic transmission method.

For the low strain reflection wave method, vibration signals are applied at the top of piles to generate stress waves, and reflection and transmission will be generated if discontinuous interface

Fig. 3.27 Bridge Pile Foundation Quality Inspection by Acoustic Transmission Method

(such as reduced sections, hole enlargement, honeycombs, mud inclusion, fractures, cavities, other defects and pile bottom) is encountered when stress waves transmit through pile bodies. The transmission time and wave shape of reflection waves in pile bodies are recorded by instruments, the integrity and defect positions of pile bodies may be determined through analyzing the characteristics of reflection wave curves, and the concrete quality of pile bodies may be estimated.

For the acoustic transmission method (see Fig. 3.27), three or more acoustic testing pipes are embedded in a pile, and the lower end of pipes is sealed. During testing, pipes are filled with fresh water as the couplant, and transmitting and receiving transducers are placed in two pipes respectively. The two transducers are placed at a same level or have a certain height difference, and they are synchronously lifted or lowered along acoustic testing pipes to obtain the variation curve of wave transit time or wave velocity along pile length. If the quality of concrete encountered in the transmission route is poor (for example, having separation, mud inclusion and other defects), acoustic waves will attenuate, and some waves will bypass defects and keep transmitting, so the transmission time will be lengthened and the wave velocity will be reduced (i.e. the diffraction phenomenon). If air interfaces with cavities are encountered, reflection and scattering will be generated to reduce the wave amplitude. That is to say, the concrete-pouring quality of pile foundation is determined qualitatively according to the distortion phenomena of acoustic wave shapes.

(2) The quantitative experiment method refers to an experiment form that measures numerical values of projects or structure, or determines whether some factors are related. For example:

Railways in coastal areas normally have such geological characteristics as deeply buried bedrock, shallow underground water level and deep soft soil at the surface layer, therefore, the pile foundation is usually adopted. The bearing characteristics of pile foundation are related to the comfort degree and safety of trains as well as quantities of repair and maintenance work of railway lines. Soft soil is characterized by great void ratio, high water content, high compressibility and long deformation time, so its engineering geological conditions are very unfavorable to bridge works. The physico-mechanical indicators of subsoil as well as various parameters of foundation design greatly affect the design and construction of bridge foundations. In order to guarantee the construction period and quality of railway projects, it is necessary to inspect design parameters and indicators of pile foundation, verify and optimize shop drawings, and improve design reliability. During railway engineering consultation, the significance of on-site pile foundation tests to the optimization of pile foundation design is specially emphasized. The proportionality coefficients (m and m_0) limit frictional resistance around pile (f_i) and other parameters in the subsoil coefficients of *Code for Design of Subgrade and Foundation of Railway Bridges and Culverts* are reasonably modified as per pile foundation test results to optimize pile foundation design and verify settlement calculation.

A static load test for vertical compression of a single pile is carried out by using anchor pile beam counterforce devices to determine the curve of relation among a single pile's ultimate vertical bearing capacity, static load and top displacement, to determine the curve of relation between pile top displacement and time under a constant load, and to provide basis for the design of engineering piles; a horizontal static load test for a single pile is carried out by using the horizontal thrust provided by anchor piles to obtain the transmission law of horizontal force of test piles and determine the proportionality coefficient (m) of a single pile's ultimate horizontal bearing capacity and subsoil horizontal resistance coefficient. Tests on pile body stress and pile end resistance are carried out to understand the load transmission law of piles as well as determine the pile side frictional resistance in different layers and pile end bearing capacity. These tests are performed by embedding reinforcement stress detectors in pile bodies and embedding pressure boxes at pile ends.

(3) The factorial experiment method refers to the tests aiming at finding out causes through known results. In railway projects, the causes of existing problems are normally identified or determined through tests. For example:

During the prefabrication of prestressed concrete box girders for a railway (Fig. 3.29), after the concrete pouring and the initial tensioning of prestressed steel strands, it is identified that there are cracks near the centerline of a box girder at the bottom of the top plate and they spread along the longitudinal direction of the box girder. Through further observation, these cracks are obviously opened during the day and closed during the night. In order to determine causes of the cracks, it is necessary to carry out the following four tests, including: (1) a temperature difference test for the top plate of the box girder, i.e. placing temperature sensors at the top and bottom surface of the top plate of the box girder, and using infrared thermometers to measure the ambient temperature at the top and bottom surface of the top plate; (2) a longitudinal loading test, i.e. placing strain measurement points at the bottom surface of bridge deck slabs for longitudinal and transverse steel strings, and placing loading points on bridge deck slabs along the longitudinal direction of the bridge to measure the stress at the bottom surface of bridge deck slabs; (3) a transverse loading test, i.e. placing loading points on bridge deck slabs along the transverse direction of the bridge to determine the influence of transverse load distribution of bridge deck slabs against the stress of bridge deck slabs; (4) a crack width test. Figure 3.28 shows a static load test for vertical compression of a single pile.

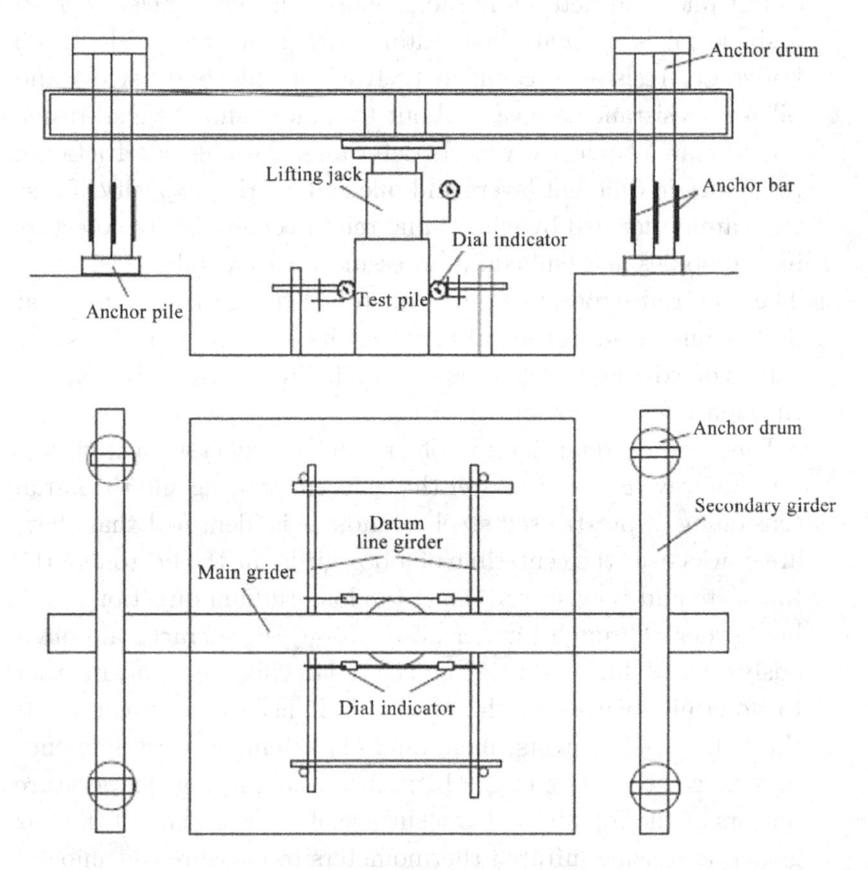

Fig. 3.28 Static Load Test for Vertical Compression of a Single Pile

Fig. 3.29 Trial Fabrication of Prestressed Concrete Box Girders for Railways

The factorial experiment indicates that the ambient temperature difference between the top and bottom surface of the top plate of the box girder is 15–20°C during sunny days. The top surface of the top plate reaches the lowest and highest temperature at 6 a.m. and 4 p.m. respectively, and the maximum positive temperature difference appears during 3–4 p.m. The ambient temperature difference between the top and bottom surface of the top plate of the box girder is 5–10°C during cloudy days. Under the actual load, the longitudinal continuity of bridge deck slabs is good, and the variation trend of the longitudinal stress influence line is basically the same with the theoretical calculation result. However, the longitudinal cracks of bridge deck slabs cause transverse stiffness reduction and make the measured values become greater than the theoretical calculation values. Under the actual load, the transverse strain of bridge deck slabs is obviously not continuous and the measured values in some degree correspond to the longitudinal crack degree and direction of bridge deck slabs at corresponding positions. According to the test load and actual crack width, it is calculated that the crack width is 0.01 mm when only the dead load takes effect. The influence of longitudinal cracks of bridge deck slabs against structure stress is small, and these cracks mainly influence the durability of the structure.

(4) The comparison experiment method refers to that: for two or more structures, one structure is used for comparison and its technical parameters remain unchanged, while the other one is used as the experimental group, and its indicators are measured after changing its certain technical parameters and then compared with the test indicators of the structure whose technical parameters are not changed to determine the influences of parameter changes against projects. For example:

The completely weathered granite layer is widely distributed in southeast China, and it has poor physical and mechanical properties, a small cohesive force, a loose structure, a low bearing capacity, a high liquid-plastic limit and relatively high silt content. It also has a low strength indicator, a low California Bearing Ratio (CBR) value, a small modulus of resilience and a poor stability against deformation and water. Meanwhile, its loose structure causes difficulties in sampling and testing, so it is hard to reflect its real strength and deformation indicators through conventional indoor tests. These areas often lack other filling material, and the completely weathered granite may be used as the filling material for the part below subgrade bed after improvement. Not only does this save project investment, it reduces spoils and protects the environment, and makes considerable economic and social benefits. In order to test the improvement effects, the base course of subgrade bed in cutting sections filled with the completely weathered granite will be directly used as the comparison group, and the base course of subgrade bed in cutting sections filled with completely weathered granite, cement and quicklime will be used as the experimental group, and the filling technologies, quality test methods, control standards, settlement and deformation characteristics, water stability and subgrade bed dynamic stress characteristics of the two groups are compared through tests.

(5) The intermediate experiment method refers to the tests used for inspecting design schemes and making preparation for project construction. These tests are carried out to determine whether design schemes are safe, reliable, technically advanced and economically reasonable, and to reveal the problems of design

Fig. 3.30 Tests for Simply-supported Box Girders

schemes as well as modify and correct the problems, so that design schemes may be used in projects. For example:

Before the batch production of common-span prestressed concrete simply-supported box girders, tests are carried out for box girder prefabrication in girder fabrication yards (see Fig. 3.30). Firstly, tests for prefabrication technologies are carried out, including: (i) the test for the development of concrete's hydration heating temperature along time, (ii) the test for transient loss of various prestresses (anchor openings, bell mouths and pipelines), (iii) the test for the prestress transmission length, (iv) the test for the prestressing effect and elastic camber upon final tensioning, (v) the test for local stress at girder ends under prestressed conditions and (vi) the test for the overall and local stress of box girders during transport. Through these tests, the mechanical properties and durability of concrete mixture (before and after hardening), the concrete mix proportion, the construction methods and technical requirements of concrete, the reasonability of pouring technologies, the material and technologies for pouring between the lower bearing plates and bearing pad stones, the reasonability of construction quality and fabrication technologies for the prefabrication of box girders, and the technical requirements on transport and erection of box girders. Secondly, the stress and deformation test is carried out under the design load to verify the design parameters of girder bodies and inspect whether mechanical properties, deformation and performance indicators of girder bodies are able to meet the design and operation requirements. Thirdly, the crack resistance test (including stiffness and crack resistance tests for girder bodies) is carried out to inspect the crack resistance performance of girder bodies. Fourthly, cracking tests and severe cracking tests (including the determination of cracking load and severe crack load of box girders as well as inspections for cracking and failure forms of box girders) are carried out to inspect the anti-cracking safety of girder bodies as well as determine the occurrence and distribution patterns of cracks. The above tests are carried out to determine key processes, verify design parameters, complete technical conditions and make sure large-scale production is feasible.

3.5.3 *Characteristics and precautions*

(1) The problems in railway projects are connected to each other in a complicated way and are affected by various accidental factors. Some characteristics are covered up or interfered in by other factors, so they may not be observed. The experiment method is adopted to isolate issues from complicated relations, to exclude subordinate, accidental and affiliated factors, and to reveal processes that cause problems with a simplified and pure state.

(2) The problems in projects are always identified after they cause consequences, so it is hard to understand and observe their development process, or observe them repeatedly. Through tests, the development process of problems may be repeatedly revealed under manual control to guarantee the accuracy of observation material and data.

(3) By using instruments and equipment during observation, oriented-strengthening is realized for physical processes under

manually controlled conditions, and special conditions that do not usually appear under normal situations are created.

(4) The generating process is very short for some phenomena and very slow for some other, but through using the experiment method and controlling test conditions, the process may be slowed down or accelerated to facilitate research.

(5) When compared with projects, tests are of a smaller scale, a shorter period and less cost, so even if tests fail, the loss is relatively smaller.

(6) The experiment method combines perceptual knowledge with rational thinking and provides objective conclusions. The required information may only be acquired through targeted tests, and necessary observations may only be obtained through elaborate plans.

(7) Technical innovations depend on human imagination and creativity, but the basic motive power is from tests and test results. The nature of things is accurately grasped in detail through tests.

(8) The experiment method is a sensitive and intuitional method for understanding things under manually controlled conditions, and only by giving full play to subjective initiative, can sensitive and experience materials be better obtained. Tests provide general conclusions through individual inspections, so specific tests results must not be idolized, blindly accepted or indiscriminately imitated.

(9) System errors and accidental errors should be distinguished and properly handled during tests to make correct judgment for test data.

(10) Variables in researches should be determined and controlled to reduce the interference of uncontrollable variables as much as possible. The design, arrangement, operation and results of tests shall be carefully inspected to avoid readily believing unreliable test results.

(11) Where possible, comparison experiments should be carried out to eliminate the interference caused by uncontrollable variables.

3.6 Simulation Method

3.6.1 *Definition*

The consulting agency simulates engineering conditions or engineering structures and fabricates models to carry out tests or establishes models to perform analysis and calculation for key technologies and complicated structures in railway projects according to the principle of similitude, then analogizes the results to actual projects to indirectly study the patterns of actual projects and understand actual projects.

3.6.2 *Classification*

According to the similar relationship between models and prototypes, the simulation method is classified into the physical simulation method and the mathematical simulation method.

(1) The physical simulation method refers to fabricating models according to a proportion similar to actual structural mechanic characteristics to carry out tests and researches. For example:

During the consultation for the wind-resistant performance of cable-stayed bridges with steel truss girders, the physical simulation method is adopted to carry out wind tunnel tests for aeroelasticity models of cable-stayed bridges with steel truss girders. See Fig. 3.31. Bridge towers and girder bodies are fabricated as per the reduced geometric scale. According to the principle of similarity on mechanical properties, the stiffness of bridge towers is provided by metal core girders, the wind effect on bridge towers is transferred to metal core girders, and the insufficient mass is provided by lead bricks located in geometric models. According to the principle of similarity on mass, steel truss girders are fabricated with lead and plastic. According to the principle of similarity on mechanical properties, the stiffness is simulated by adding special U-shaped spring connections at each segment of truss girders.

(a) Picture of Complete Model

(b) Picture of Local Part

Fig. 3.31 Wind Tunnel Test for Aeroelasticity Model of a Cable-stayed Bridge with Steel Truss Girders

(2) The mathematical simulation method is based on the mathematical and physical similarities between models and engineering conditions or engineering structures. For example:

During the consultation for the design of cross and vertical sections for lines, the design parameters of cross and vertical sections under high-speed running conditions have huge influence on the comfort and safety of trains, so it is hard to evaluate the dynamic wheel–track safety indicators (including the transverse force of wheel/track, coefficient of derailment, rate of wheel load

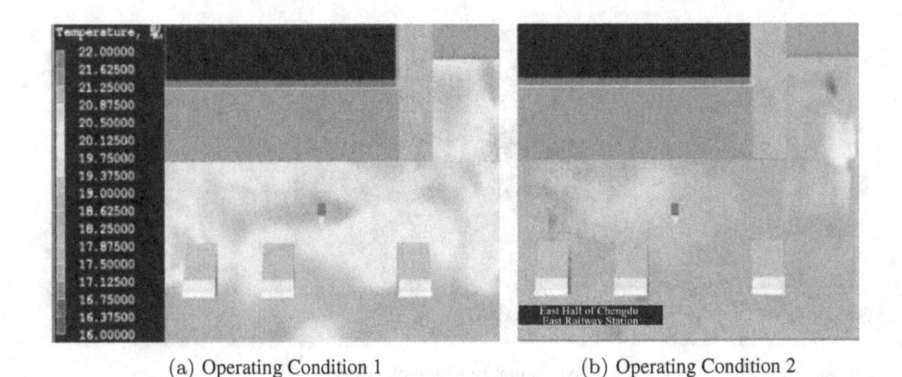

(a) Operating Condition 1 (b) Operating Condition 2

Fig. 3.32 Large Space Airflow Simulation Analysis

reduction and dynamic gauge widening) and the comfort indicators (including the vertical and transverse vibration acceleration as well as stability of trains) using conventional design methods when trains pass through plane curves and vertical curves. Therefore, it is necessary to create a coupled dynamic model for the train-line system based on the mathematical and physical similarities between the actual train and structure, input space curve parameters of cross and vertical sections of the designed line into the computer simulation system, simulate and analyze the operation safety and comfort, evaluate the reasonability of design parameters of cross and vertical sections of the line, and optimize the design of cross and vertical sections of the line.

During the consultation for the OCS, the emulation technique is adopted to carry out a dynamic simulation analysis for the OCS to propose schemes, technical conditions and technical parameters for the optimization of the OCS as well as optimization opinions on key design schemes including the suspension method of the OCS, the arrangement method for overhead crossings at turnout positions, the OCS design with neutral sections at contact line section overlaps, the design scheme of interface reservation for civil works (bridges, tunnels, subgrade, stations and yards) of the OCS, the arrangement scheme for traction feeding cables on bridges and the arrangement scheme for power supply cables crossing tracks.

During the consultation for the heating supply system, the ground-source heat pump system is adopted to carry out an annual energy consumption simulation analysis to provide optimization opinions and guarantee the safe and reliable operation of the heating supply system. According to the reliable data obtained from site tests for ground-source heat pump systems at different stations, the heat transfer analysis is performed, and the soil temperature rise must not exceed 3°C after 50 years' operation of the control system. An annual energy consumption simulation analysis is carried out for ground-source heat pump systems at different stations. A large space airflow simulation analysis is shown in Fig. 3.32.

3.6.3 *Characteristics and precautions*

(1) Models in the simulation method are the test subjects replacing engineering conditions or engineering structures, and also test approaches for engineering conditions or engineering structures. On the one hand, models are the subjects for consulting personnel to test with test tools, and on the other hand, they are only the substitutions of engineering conditions or engineering structures. The real subjects of tests are engineering conditions or engineering structures, and models are the tools for consulting personnel to understand engineering conditions or engineering structures.
(2) For the simulation method, the similarities among objective things are its objective basis, and the similarities between models and engineering conditions or engineering structures are its premise. It adopts the analogic reasoning method (the analogy method is discussed in Section 3.9) to analogize the research results of models to engineering conditions or engineering structures.
(3) Models may be different from engineering conditions or engineering structures in the nature of substance, but the similarities derived from the identity of mathematical forms are beyond the limit of physical simulation. Therefore, engineering conditions

or engineering structures may be replaced by systems that are simple, inexpensive and easy to study.

(4) The simulation method is able to enlarge or reduce engineering structures and to compress time. For example, a structure with a length of several hundred meters may be tested using a model with a length of several meters, and the degree of structure fatigue damage caused by trains in 100 years may be simulated by fatigue loading tests in over 10 days.

(5) The simulation method may also be used as the auxiliary analysis tool for consultation. For example, when analyzing the dynamic performance of bridge structures, bridge models are created by using bridge modeling software including MIDAS, Dr. Bridge and ANSYS; spatial coupled vibration models for the train system, track system and bridge system are established respectively by using MSC.PATRAN, MSC.NASTRAN and MSC.ADAMS/RAIL; the dynamic effects of trains against bridges as well as the influences of track regularity, train speed, bridge stiffness and deformation on the safety and comfort of trains are analyzed by using dynamic structural analysis software.

(6) When using the simulation method, there must be similarities between models and engineering conditions or engineering structures, in order to carry out research by using models to replace engineering conditions or engineering structures, and to acquire information about engineering conditions or engineering structures through research on models.

3.7 Question-raising Method

3.7.1 *Definition*

The question-raising method proposes questions to identify crucial reasons of design and projects and to put forward consultation opinions on optimization, completion and improvement. The most representative question-raising method is the checklist method. For the safety, quality, investment and construction period problems in existing railway projects or design, the common mistakes, errors, omissions and conflicts in design, the quality accidents and engineering diseases in existing railway projects as well as the experience and results of existing projects or design, the checklist method adopts standardized lists to list the above-mentioned items according to different disciplines and interfaces, carries out comparison, analysis, discussion and judgment by items, summarizes major problems and provides solutions.

3.7.2 *Classification*

The checklist method proposes questions on seven aspects including (1) analogy and imitation, (2) change and transformation, (3) extension and enlargement, (4) reduction and decrease, (5) substitution, (6) change of sequence and (7) combination to identify problems, imitate existing successful experience, methods and measures, optimize and improve project safety and quality, facilitate construction, control investment and optimize construction organization plans.

(1) Analogy and imitation

Analogy and imitation refer to raising questions on whether the successful cases or problems encountered in existing projects are similar to those in projects under consultation. For example:

Underground pipelines (Fig. 3.33) and cables may not be accurately positioned during surveys or their positions may not be marked out on drawings, so during railway project construction, they may be damaged due to excavation or may conflict with railway lines and bridges and cause rerouting at local sections or adjusting positions of pier and abutment of bridges. Therefore, questions are raised on whether the survey for underground pipelines and cables meets survey procedures, on the feasibility and economy of relocation and protection measures, on whether the plane layout, elevation and bore diameter of underground pipelines are marked out in route plans, line profiles and work point design drawings, so that consulting personnel is able to carry out examination and verification according to these questions.

Fig. 3.33 Underground Pipelines

Fig. 3.34 Construction of Simply-supported Full-span Box Girders with Mobile Formwork in Bridge–tunnel Connection Section (instead of Prefabrication and Erection)

(2) Change and transformation

Change and transformation refer to raising questions on whether it is able to change and transform projects under consultation. For example:

During the consultation for the construction scheme of simply-supported full-span box girders in bridge–tunnel connection sections (see Fig. 3.34), questions are raised on the economy, reasonability and construction period influences of schemes including: (i) adding a girder fabrication and storage yard to prevent transporting girders through tunnels; (ii) enlarging tunnel sections; (iii) construction of cast-*in-situ* box girder; (iv) decreasing box girder width and transporting girders to pass through tunnels with flatbed trucks; (v) adopting double-piece combined box girders instead of full-span box girders when passing through tunnels; (vi) post pouring of flange plates of full-span box girders; (vii) constructing girder transport roads avoiding tunnels; (viii) adopting combined box girders to replace full-span box girders.

For another example, opinions on adjusting relevant geological parameters in the design settlement calculation for soft soil and mollisol subgrade are proposed based on the results of subgrade preloading tests and CFG pile tests.

(3) Extension and enlargement Extension and enlargement refer to raising questions on whether engineering schemes or measures are able to expand the scope of application, add functions, extend service life and strengthen engineering measures. For example:

Generally speaking, for high-speed railways, a subgrade structure is adopted when the height from the rail surface to the ground surface is less than 8 m, while a bridge structure is adopted when that height is equal to or greater than 8 m. However, during a consultation, questions are often raised on whether it is suitable to use bridges to replace subgrade or enlarge the scope of bridges when high-speed railways are located at suburb sections (see Fig. 3.35), sections with soft soil subgrade, sections with concentrated ponds, sections with concentrated roads and ditches, sections with difficulties in subgrade settlement control, and sections with short subgrade.

(4) Reduction and decrease

Reduction and decrease refer to raising questions on whether engineering schemes or measures are able to narrow the scope of application, reduce volume, lower weight, save components and decrease quantity. For example:

Fig. 3.35 Using Bridges to Replace Subgrade for High-speed Railways at Suburbs

In plain areas, the elevation of railway lines and the length of bridges may be reduced by determining a reasonable clearance under bridges, selecting reasonable bridge types, reducing bridge span as well as adopting reasonable bridge construction methods and measures. Therefore, questions are raised on (i) improving the arrangement of grade and grade sections; (ii) verifying the class of roads or waterways under bridges; (iii) adopting through-type bridges; (iv) using oblique rigid frame bridges; (v) adopting the bridge type that pier bodies are perpendicular to roads or waterways under bridges while bend caps are perpendicular to railways; (vi) adopting the dragging or incremental launching construction scheme; (vii) adopting the rotation construction scheme; (viii) lowering the elevation of roads under bridges; and (ix) adopting temporary construction access roads and transitional measures. This is so that consulting personnel are able to carry out verification by comparing to these questions and propose consultation opinions on reducing the height of bridges below rail surface, the span of bridges and the clearance height

Fig. 3.36 Through-type Bowstring Arch Bridge Capable of Effectively Reducing Elevation of High-speed Railways

occupied by construction equipment under bridges to reduce the design length of bridges, the length of bridges and the investment of projects. For the bowstring arch bridge that has a span of 128 m and is shown in Fig. 3.36, the height of the structure below rail surface may be reduced by 6–7 m when it is compared with a continuous girder bridge with a same span.

For another example, feasible consultation opinions are proposed on reducing the adjustment value of rail expansion or avoiding the setting of rail expansion joints based on optimizing the structural design of large-span bridges and selecting reasonable fastener types and fastening forces. Consultation opinions on reducing the scale and quantity of prefabrication yards for girders are proposed according to construction organization plans of railway lines, structural types of simply-supported girders, arrangement and quantity of bridge spans as well as construction schemes.

(5) Substitution

Substitution refers to raising questions on whether the structural types, layout patterns, material and construction methods of project under consultation are able to be replaced by other ones.

For example:

During the consultation for a high rock cutting, questions are raised on whether it is necessary to change cuttings into open cut tunnels or shed tunnel, on whether it is necessary to set passive rockfall barriers, rockfall platforms and rockfall channels, and on whether there are other surface protection methods for slopes above railway lines after analyzing whether slope protection and reinforcement measures are able to effectively intercept small rocks falling from slope surface, and whether the strength of protective structures is able to withstand the impact of large rocks fall from slope surface (see Fig. 3.37).

(6) Change of sequence and position

Change of sequence and position refers to raising questions on whether it is able to change layout patterns, construction sequences and positions of projects under consultation. For example:

During the consultation for girder-type continuous structures, questions are raised on the construction sequence of side span closure first and mid-span closure second as well as the construction sequence of mid-span closure first and side span closure second according to projects' landform, traffic, hydrology, geology, ration of side span and mid-span, structural stress, structure stability, project investment, bearing settlement and other engineering conditions, so that consulting personnel is able to carry out examination and verification according to these questions.

For another example, questions are raised on the adjustment of relative positions based on the principle of comprehensively considering functions of stations and sections and meeting these functions to achieve a reasonable and compact layout as well as unobstructed roads, save land and meet landscape requirements at station areas, which is shown in Fig. 3.38.

(7) Combination

Combination refers to raising questions on whether it is possible to combine several engineering components or combining

Fig. 3.37 Setting Rockfall Barriers for Dangerous Rocks on Slope Surface at Both Sides of Railways

several engineering layout patterns under consultation. For example:

It is necessary to set dampers for continuous girder bridges in high intensity earthquake zones (see Fig. 3.39). The vertical force of bridges is born by two bearings at each pier. The tonnage of dampers at movable bearings is generally determined according to the earthquake force calculation, embedded parts are installed on girder bottom and piers when pouring girder bodies and piers, and dampers are installed on embedded parts upon completion

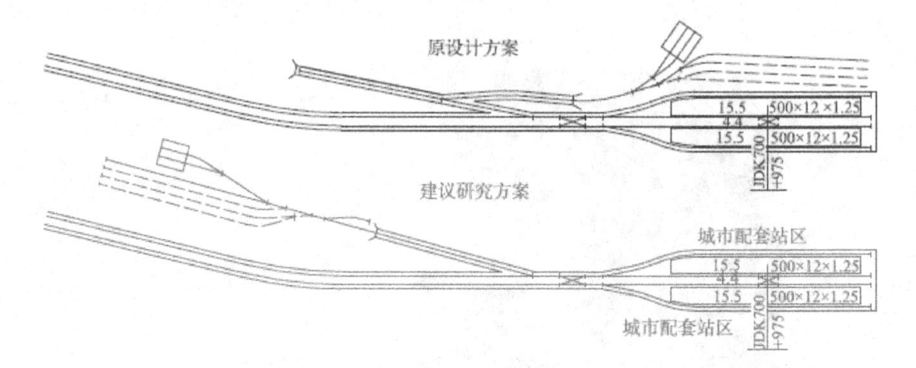

Fig. 3.38 Optimization for Plane Layout of Stations and Yards

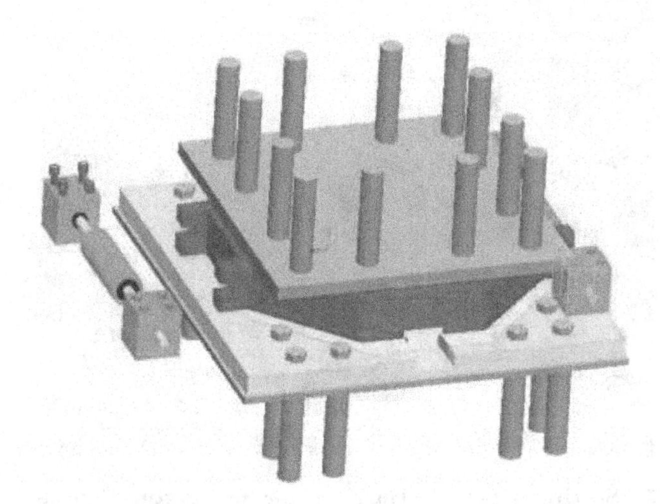

Fig. 3.39 Three Dimensional Diagram of Pot-type Rubber Bearing with Dampers

of girder erection. Each pier is provided with one damp with a great tonnage and a large volume, so the installation is difficult and the bridge appearance is not good. During consultation, questions are raised on whether it is possible to divide a damper into four parts, to reduce the damper dimension and make a damper become part of a bearing, and to install dampers at both sides of each bearing in factory (i.e. combining bearings with dampers), so that the bridge appearance may be improved, and the damper installation process may be skipped during construction to accelerate construction progress.

3.8 Induction Method

3.8.1 *Definition*

The induction method refers to consulting personnel summarizing common problems in work points, engineering measures and design parameters of projects or design under consultation through site verification, design document review, tests and researches (when necessary) and model tests (when necessary), and then obtain evaluations on projects or design documents as well as guiding opinions on optimization, improvement and completion. For example:

The spheroidal weathering is a significant unfavorable geological phenomenon in granite sections, and without sufficient understanding of its distribution characteristics as well as improvement in engineering survey measures, problems including pile breaking, construction cost increase, uneven settlement and structural instability may occur during project construction and operation. Considering the distribution of weathered spheres is characterized by serious discreteness, deep buried depth and irregular spatial occurrence characteristics, it is difficult to accurately reveal their actual shape. Firstly, from the angle of their formation mechanism, it is known that weathered spheres are of an elliptical shape. The diameter or maximum length cannot be revealed by drilling along the longitudinal length of weathered spheres, but because a same survey method is adopted, the longitudinal length revealed by drilling may be used as the relative height or statistical diameter of weathered spheres to reflect their size variation characteristics. Through relevant analyses on the size and elevation of weathered spheres, the relation formula between sphere diameter and elevation as well as the correlation on overall trend (i.e. the trend that the size of weathered spheres reduces when the elevation increases) are summarized as shown in Fig. 3.40. Supplementary geophysical prospecting means, standards and technical requirements on the spheroidal weathering in granite sections may be further summarized.

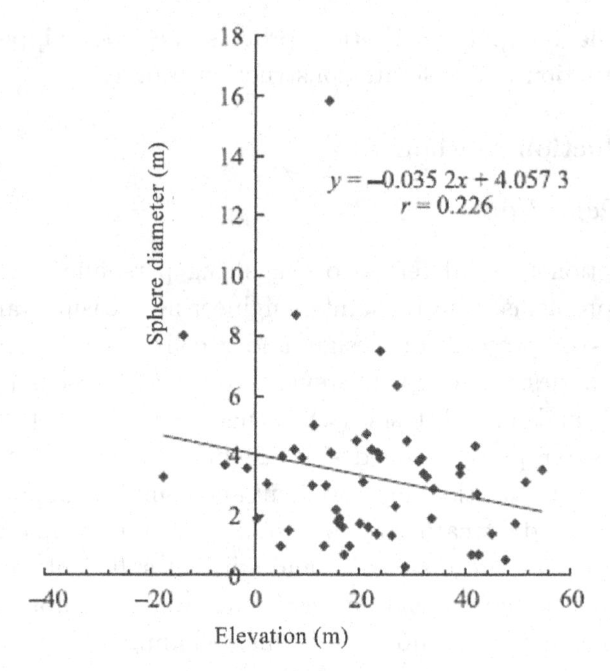

Fig. 3.40 Relationship between Size and Distribution Elevation of Weathered Spheres in Subgrade Section of a Railway

3.8.2 *Classification*

According to the completeness of induction premises, the induction method is classified into the complete induction method and the incomplete induction method. The complete induction method obtains general conclusions from a certain property or relation of all objects included in premises according to sufficient reasons. The incomplete induction method is classified into the simple enumeration induction method and the scientific induction method as per whether the law of causality is adopted. The simple enumeration induction method is a basic type of the incomplete induction method, and it makes a general conclusion that a certain type of objects are of a certain property because part of the objects are of this property. The scientific induction method is adopted in railway engineering consultation, and it makes a general conclusion that a certain type of projects or engineering problems are of a certain property based on the understanding of some properties as well as the necessary connections between properties of a certain railway project or problems in the project.

The scientific induction method includes the method of agreement, the method of difference, the joint method of agreement and difference, the method of concomitant variation and the method of residues.

(1) Method of agreement

In this method, if a same problem investigated in different occasions only has one associated condition in common, then this common condition is in a causal association with the problem. For example:

Through investigations and analyses of culverts that are used for flood discharge and traffic purposes which have different hole diameters, lengths, foundation dimensions and filling depths and encounter impeded drainage at outlets, it is identified that their foundation is located on soft subsoil, so it can be concluded that there is a causal relationship between placing culvert foundation on soft subsoil and impeded drainage at culvert outlets (see Fig. 3.41).

(2) Method of difference

This method compares two occasions, where one of them has a problem, while the other does not, and if there is only one different condition in the two occasions, then this condition is the cause of the problem. For example:

Through analyses of culverts with their foundation on soft subsoil, it is known that the impeded drainage will not occur if subgrade treatment (including replacement, CFG piles and rotary jet piles) is performed for these culverts, it can be concluded that the cause of impeded drainage is placing culvert foundation on soft subsoil without foundation treatment.

(3) Joint method of agreement and difference

In this method, if a common condition exists in several occasions that have a same problem, while the common condition does not exist in several other occasions that do not have the

Fig. 3.41 Drainage Ditch at Culvert Outlet

problem, then this common condition is the cause of the problem. For example:

If culverts with impeded drainage at their outlets are all located on soft subsoil, while culverts without impeded drainage at their outlets are not located on soft subsoil, so it can be concluded that the cause of impeded drainage is placing culvert foundation on subsoil.

(4) Method of concomitant variation

In this method, if a problem changes in different occasions when only one condition changes, then this condition is in a causal association with the problem. For example:

Through analyses of culverts that are used for flood discharge and traffic purposes which have different hole diameters, lengths, foundation dimensions and filling depths and have soft soil and non-soft subsoil, it is identified that the impeded drainage only occurs when the foundation is on soft subsoil, so it can be concluded that there is a causal relationship between the soft subsoil and impeded drainage.

(5) Method of residues

The method of residues is also known as the dissection method or the exclusive method, and it refers to a situation where if a complicated problem A is caused by another complicated cause B, and in the meantime, it is known that a part of the problem A is caused by a part of the cause B, then the rest of the problem A is caused by the rest of the cause B. For example:

For culverts, the uneven settlement, impeded drainage at outlets and insufficient discharge capacity are caused by soft subsoil, poor structural integrity and small size, and it is known that the uneven settlement and impeded drainage at outlets are caused by soft subsoil and poor structural integrity, therefore, the small size is the cause of insufficient discharge capacity.

3.8.3 *Precautions*

The conclusions of the above-mentioned methods are of a probability, so only using these methods for consultation is not enough. For example, in the residual method, the cause of insufficient discharge capacity of converts may be related to the settlement of culverts on soft subsoil, while it may also be related to the wrong elevation of culverts or culvert outlets. Therefore, correct consultation conclusions may only be obtained from complicated factors by using multiple methods comprehensively, grasping engineering technologies skillfully and understanding actual project conditions accurately.

3.9 Analogy Method

3.9.1 *Definition*

The analogy method refers to consulting derived from corresponding consultation conclusions and solutions for problems identified during engineering consultation based on existing similar projects and problems in these project, or according to engineering tests, model tests and model analyses carried out during consultation.

Analogy refers to comparisons between similarities of new facts and known objects. The analogy method is an effective and tentative judgment method for a further understanding of things based on their forms, properties, structures and functions, the principles, forms, means and contents of theories as well as the existing knowledge.

3.9.2 *Classification*

The analogy method is classified into (1) the simple coexistence analogy method, (2) the causal relationship analogy method, (3) the symmetrical relationship analogy method and (4) the comprehensive analogy method.

(1) Simple coexistence analogy method

The simple coexistence analogy method derives conclusions according to the similarities of superficial phenomena between projects under consultation (or problems in these projects) and existing projects (or problems in these projects), so these conclusions are subjective to a great extent. For example:

It is known that in the design of existing large-span prestressed concrete rigid frame continuous bridges, when piers are relatively short, piers with a relatively large stiffness will greatly restrain the deformation of girder bodies, and the temperature effect will cause the top of the pier to undergo a great bending moment, so the design of piers is often very difficult. In order to reduce the structural temperature effect, some piers fixed to girders are often changed into piers with movable bearings, so bridges become prestressed concrete rigid frame continuous bridges (see Fig. 3.42).

For elevated stations with combined bridges and buildings, if multi-column prestressed concrete bent cap frame piers are adopted, used for supporting main track girders, station platform girders and are even used as the foundation of roof truss columns of stations for supporting roof truss columns, the length of bent caps for frame piers will be too great. So according to the

Fig. 3.42 Prestressed Concrete Long Bent Cap Frame Pier with Some Columns Provided with Movable Bearings at Their Top

superficial similarities between large-span prestressed concrete rigid frame continuous bridges and multi-column prestressed concrete long bent cap frame piers, it is derived by using the simple coexistence method that frame piers are relatively low in height, and the temperature effect will cause the top of the column to undergo a great bending moment, so the design of frame piers will be very difficult and the structural stress is very unfavorable. Changing some columns (of frame piers) fixed to bent caps into columns with movable bearing at their top may effectively reduce the structural temperature effect and improve the structural stress.

(2) Causal relationship analogy method

The causal relationship analogy method derives conclusions according to the same causal relationship between each property of projects under consultation (or problems in these projects) and existing projects (or problems in these projects), so these conclusions are of a higher reliability but still a relatively great probability. For example:

In case the vertical inherent frequency of bridges is close to the vertical vibration frequency when trains pass through bridges with a certain speed, bridge resonance will occur and cause a peak dynamic coefficient, and as a result, violent vibration, ballast looseness and rail damage will occur, the normal operation of track structure will be affected, concrete cracking, structure fatigue and bearing capacity decrease will also occur, and even the safety of bridges will be endangered. Along with the increase of inherent frequency of bridges, the general trend of dynamic coefficient is decreasing but not monotonously. The dynamic coefficient will decrease along with the increase of bridge stiffness. In order to avoid bridge resonance and reduce dynamic effects, different limiting values for the minimum inherent frequency of railway bridges may be formulated as per differences in operation speed.

For subgrade and bridges sections provided with acoustic barriers, air in the line space will inevitably be pushed aside by trains when they pass through these sections with a high speed, and an obvious pressure fluctuation will be generated on acoustic barriers (see Fig. 3.43). In case the natural vibration frequency of acoustic barriers becomes the same with the dynamic response frequency of fluctuating wind generated by high-speed trains, a resonance response being much greater than the static force effect will be generated, amplifying the response of fluctuating wind generated by trains to acoustic barriers by several times and causing accumulated damages, dropping and destruction of acoustic barriers. According to the same causal relationship

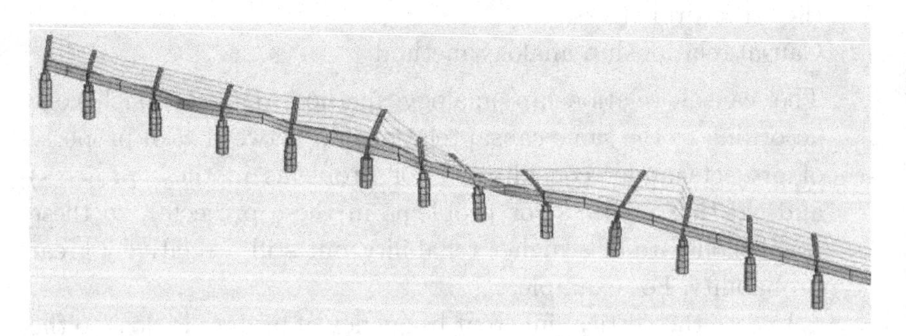

Fig. 3.43 Structural Deformation Shapes of Acoustic Barriers under Fluctuating Force Effect of High-speed Railways

between the bridge resonance caused by the vertical impact effect of trains and the acoustic barrier resonance caused by aerodynamic effects, it is derived by using the causal relationship method that different limiting values for the minimum inherent frequency should be selected for acoustic barriers of high-speed railways as per differences in the operation speed of railways, and the stiffness of acoustic barriers should be increased to reduce the fluctuating force coefficient.

(3) Symmetrical relationship analogy method

The symmetrical relationship analogy method derives conclusions according to the symmetrical relationship between the properties of projects under consultation (or problems in these projects). For example:

For high-speed bridges with ballastless tracks, the track smoothness is in a symmetrical relationship with the train speed. The better the track smoothness is, the faster the trains are able to run on the tracks. In case the smoothness requirements on trains passing through with a speed of 300 km/h are met when the mid-span deflection of 32 m simply-supported girders is not greater than 10 mm, it is derived by using the symmetrical relationship analogy method that the limiting value for the mid-span deflection of 32 m simply-supported girders should be more strict so as to meet the smoothness requirements on trains passing through with a speed of 350 km/h. By referring to *General Technical Specifications and Circulars of Ballastless Tracks* (DB Netz NST on August, 1, 2002) and *DIN Technical Reports 101* of Germany, it is known through further analyses that the limiting value for the mid-span deflection of 32 m simply-supported girders should not be greater than 7.5 mm so as to meet the smoothness requirements on trains passing through with a speed of 350 km/h. Figure 3.44 shows a schematic diagram for double-track full-span prestressed box girders.

(4) Comprehensive analogy method

The comprehensive analogy method is carried out according to physical or mathematical models established with a comprehensive method as well as similarities between engineering

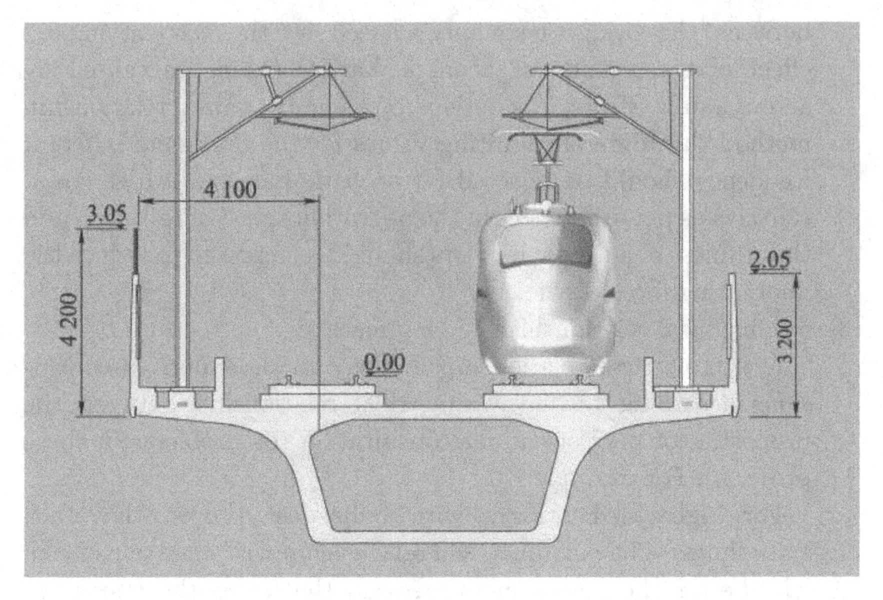

Fig. 3.44 Schematic Diagram for Double-track Full-span Prestressed Box Girders of High-speed Railways with Ballastless Tracks (Unit: mm)

conditions (or actual projects) and the aforesaid physical or mathematical models. Similarities among multiple properties of different objects are considered in the comprehensive analogy method, so the reliability of conclusions is substantially increased. However, it is impossible to consider all properties of objects, so conclusions must not be completely believed. The simulation method focuses on how to simulate engineering conditions or engineering structures, while the comprehensive analogy method emphasizes deriving consultation conclusions for engineering conditions or actual projects according to similarities between engineering conditions (or actual projects) and physical or mathematical models. For example:

For high-speed railways' through-type bowstring arch bridges with steel bridge decks, their tie girders and bridge decks are tied to arch positions of cross girders for assembly, two welds are located at both sides of bridge deck slabs, and high-strength bolts are used for assembly at both sides of cross girders. In order to

Fig. 3.45 Model of Complete Steel Bridge Deck Section for High-speed Railways

understand stress distribution rules and force transmission ways of structures during operation as well as influences of installation and welding procedures and processes against structures during operation, and to provide bases for the optimization of design and manufacturing processes, a section model test is carried out using one section as per the proportion of 1:2 between the model and the actual structure (see Fig. 3.45), i.e. the simulation is performed as per the proportion of 1:2 between the model (consisting of tie girders, main cross girders, secondary cross girders, longitudinal girders, bridge deck slabs, U-shaped ribs of bridge deck slabs and slab ribs of bridge deck slabs) and the actual structure. The material of the model is the same as that of the actual structure. Test loads simulate functions of single-track middle-live loads, double-track middle-live loads and secondary dead loads.

Through the test results, it is identified that the stress concentration phenomenon is obvious at arch positions of cross girders of bridge deck slabs due to the different characteristics of sections at both sides of cross girders as well as the influences of welding residual stress and welding defects of bridge deck slabs. The manufacturing, installing and load applying of models are carried out by using the comprehensive analogy method according to the similarities in physical properties between models and actual projects, and it is derived by using the comprehensive analogy method that excessive residual stress and deformation may occur at joints of cross girders and bridge deck slabs during the welding and assembly of actual bridges, so it is necessary to improve and complete technologies during welding and assembly to prevent excessive residual stress and deformation in actual engineering structures.

3.9.3 *Characteristics and precautions*

(1) The basic links of the analogy method are association and comparison. Firstly, analyses are carried out by selecting existing projects or their existing problems in analogy with projects under consultation, or by combining engineering tests, model tests and models in the consultation. Then, engineering characteristics, structural forms, functions and design methods of existing projects (or their existing problems) and projects under consultation are associated and compared, and project under consultation will be judged according to the judgment on existing projects (or their existing problems) or engineering tests, model tests and model analyses.
(2) The objective basis of analogical reasoning is the inherent universal connection among things. Unknown intermediate processes or links are omitted during analogy and the judgment on a certain project is derived directly based on another project, so there is a logical probability.
(3) The analogy method is characterized by comparison first and derivation second. The comparison is the basis of analogy, and both similarities and differences are compared. Similarities

among objects are the premise for implementing the analogy method, so it is not possible to carry out analogical reasoning for objects without similarities.

(4) The function of the analogy method proceeds from one point to another. If the first point is deemed as the premise and the next point is deemed as the conclusion, the process of analogical thinking is a reasoning process. If more and more similarities are identified between two objects under comparison and it is known that one of the two is of a certain property while this property has been identified in the other object, then it is reasonable to analogize and affirm that the other object should also have this property. Induction and deduction processes during analogy make it possible for analogy to proceed from one point to another.

(5) During analogy, as many common properties or characteristics between some objects and other similar objects should be found out as possible, correct theories should be used for guidance to find out necessary connections between known properties of objects and properties derived through analogy, and strict logical rules should be followed.

(6) The analogy method belongs to a parallel thinking method when compared with other thinking methods, so it should be carried out at the same level.

(7) The analogical reasoning is a kind of probable reasoning, so conclusions may not be true even if premises are true. In order to increase the reliability of conclusions of analogy, it is necessary to confirm as many similarities as possible between objects. The more the similarities there are, the more reliable the conclusions will be, and that is because the more similarities there are between objects, the more closely the objects are connected. On the contrary, the conclusions will be less reliable. In addition, it must be noted that the same properties in analogy premises and the derived properties should be essential. If a specific property or occasional property of a certain object is forcibly analogized to another object, it will be a mistake of improper analogy or mechanical analogy.

3.10 Deduction Method

3.10.1 *Definition*

The deduction method refers to the consulting personnel deriving principles and methods to be adopted for projects under consultation according to the universal principles, experience and methods of existing projects. In a deductive argument, universal principles, experience and methods are bases, while principles and methods to be adopted for projects under consultation are theses.

3.10.2 *Classification*

The main form of deductive reasoning is syllogism, i.e. major premises, minor premises and conclusions. Major premises are general principles, minor premises are individual things under argument and conclusions are theses. The form of deductive reasoning must be met if the deduction method is used for argument.

For example, high cutting slopes are poorer in stability than low cutting slopes, so slope collapse may be caused due to large rainfall and long-time water immersion, and rockfall may be caused due to flowing water. Therefore, if projects under consultation are located in southern areas with large rainfall, a relatively high cutting excavation height and insufficient protection measures for cutting slopes, it is derived through deductive reasoning that the cutting has potential safety hazards including slope collapse and rockfall, so cutting slopes should be reinforced and a comparison between using open cut tunnels or shed tunnels should be carried out (see Fig. 3.46). Meanwhile, according to specific conditions, it should be analyzed whether it is necessary to set passive rockfall barriers, rockfall platforms and rockfall channels for dangerous rock at the surface of slopes above railway lines.

3.10.3 *Characteristics and precautions*

(1) When using deductive reasoning, major premises as the bases must be correct and have necessary connections with conclusions,

Fig. 3.46 Cutting Slope Protection for High-speed Railways

and they must not be farfetched or detached, or else the correctness of conclusions will be doubted.

(2) For the deduction method, there are necessary connections between the universality of premises and the individuality of conclusions, the latter are contained in premises and do not exceed the scope of premises. The correctness of conclusions in the deduction method depends on whether the universality as the starting point correctly reflects actual project conditions as well as whether the connections between the existing principles, experience and methods and the specific projects under consultation are correctly reflected by the connections between premises and conclusions. If premises are universal principles, experience and methods that have been tested through practice and are able to correctly reflect actual project conditions and logical rules are followed during deduction, the conclusions acquired are reliable. The direction of movement of thoughts in the deduction method is from general to individual as well as from abstract to specific. That is to say, the premises of deduction are abstract universal principles, experience and methods, while the conclusions are specific.

(3) The deduction method is an important tool for logical proof. Deduction is a necessary thought process and if the movement of thought is logical, the conclusions are determined by premises. Therefore, selecting undoubted and reliable propositions as premises can convincingly prove or contradict certain propositions. The deduction method applies general railway engineering theories and technologies to specific projects during consultation and acquires correct deductions, so it is an important thinking method for consultation.

3.11 Systematic Method

3.11.1 *Definition*

The systematic method utilizes systematic and scientific theories and methods to study and solve various engineering problems encountered during consultation, and its application level is systematic and scientific. The key points in the method include exploring factors and processes of projects as well as optimizing engineering design. According to systematic and scientific opinions and theories, projects under consultation are deemed as a complete object to carry out overall control, interactions and change rules between factors and factors, between factors and systems, between systems and environment as well as between systems and systems are explored in a comprehensive way, and the relationship between the internal and external environment of projects are grasped.

3.11.2 *Principle*

The integrity principle, dynamic principle, optimization principle and modeling principle should be grasped when using the systematic method.

(1) The integrity principle refers to starting from an entire object, analyzing different parts and the relationship among these parts based on the entire object, and achieving a deep understanding of the entire object according to analyses of different parts. The

integrity principle is the first principle and basis of the systematic method. It deems a research object as an organic whole, explores the regularity in its composition, structure, function, motion and change, understands projects under consultation through connections and interactions, and achieves coordination in systematic analysis and system integration, induction and deduction, partial and whole as well as individual and general. For example:

During railway engineering consultation, structures including subgrade, bridges, culverts, tunnels and tracks are deemed as an organic whole to review its coordination and uniformity and guarantee the combination of trains, lines, bridges (or subgrade and tunnels) has good dynamic characteristics.

Review whether route schemes are economic and reasonable, and whether they meet urban and railway planning and development requirements and are technically feasible, and especially whether they fit urban planning. Review whether the design of horizontal and longitudinal sections of lines is reasonable and meets provisions of relevant specifications and technical standards. Review the setting of grade and slope sections as well as the optimization and adjustment for the design of horizontal and longitudinal sections according to operation conditions, interchanges, waterways and hydrological conditions.

Review whether the design of horizontal and longitudinal sections of lines coordinates with retaining works of bridges and subgrade and especially whether control conditions and corresponding engineering measures for side slope height of subgrade with high filling and deep excavation along entire lines meet relevant provisions and requirements.

Review the design of transition between subgrade and other works (bridges, tunnels and culverts, etc.) as well as the design of transitional sections between embankment and cutting and between ballast tracks and ballastless tracks.

Review the interface between the station and yard discipline and other disciplines as well as the interface between stations (and yards) and operation equipment of E&M works. Review

Fig. 3.47 Large Railway Marshaling Station

whether the design of throat areas is able to meet the receiving (departure) of trains from (or to) different directions as well as the intensive receiving (departure) function and tracking interval function of trains from (or to) a same direction. Review the design's adaptability and reasonability on avoiding cross interferences of operating in stations. A large railway marshaling station is shown in Fig. 3.47.

Analyze the actual functional requirements of train dispatching and telephone dispatching subsystems, study the functional composition of operation dispatching systems, and review the overall composition of system equipment, interface relations among subsystems as well as specific software and hardware configuration of equipment.

Review whether there is any omission in interfaces of different disciplines (including route, track, subgrade, bridge, tunnel, EMU, communication, signal, power, electrification, lighting, operation and maintenance, etc.) and whether the interface treatment is reasonable.

(2) The dynamic principle refers to that the system is always dynamic and under motion and change. Different development

stages of the system should be researched as a whole to grasp the development process and future trend. The dynamic principle is also the foundation of the systematic method. For example:

For infrastructure of lines as well as structure and equipment being hard to reconstruct, review whether conditions are reserved for long-term social, economic and railway development, whether the position, layout and scale of stations meet demands on economy, urban planning, passenger traffic volume, railway traffic organization, engineering conditions, traffic capacity and technical operation of cities along lines, and whether environmentally sensitive areas including natural reserves, scenic spots, drinking water source protection areas and key cultural relics sites under state protection are avoided during route and site selection; review whether suitable speed values as well as measures against noise and vibration are adopted when passing through cities or residential areas, and whether national environmental protection standards and requirements are met; review whether protection measures by combining green plants with projects are adopted for subgrade and side slope to meet requirements on appearance, environmental protection and water protection as well as improve railway quality.

Figure 3.48 shows the specific example of using the dynamic principle in an actual project. Acoustic barriers of an intercity railway bridge adopt a combined structure using reinforced concrete baffles and inserted acoustic barrier plates. For existing environmental sensitive points, sound-absorbing concrete slabs are installed at the inner side of baffles, and at baffle top, steel plates are embedded and semi-inserted acoustic barrier plates are installed to meet environmental protection requirements. Considering the demands of urban–rural integration development and social development, steel plates are also embedded at baffle top for locations that are currently not environmentally sensitive as the basis for adding inserted acoustic barriers in the future.

(3) The optimization principle or the overall optimization principle is the purpose and requirement of using the systematic method. It is required to make overall arrangement and coordination, select

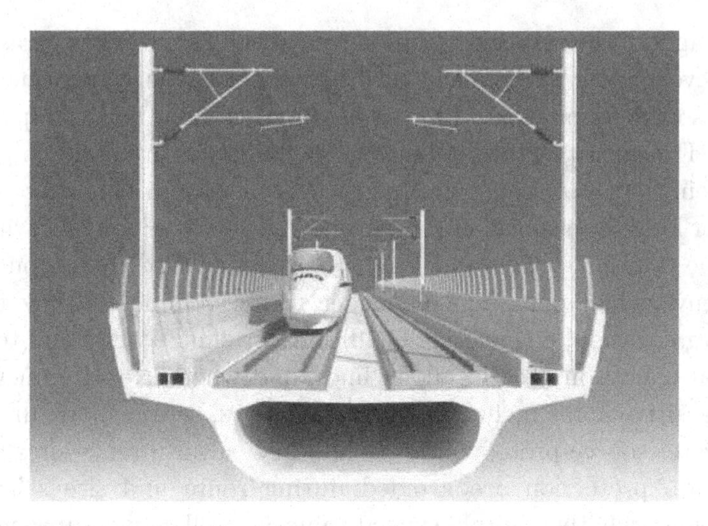

Fig. 3.48 Effect Drawing of Intercity Railway Girder Considering Social Development

the best among various objects, and adopt the peak value and optimal point of time, space, procedures, subjects and objects to conduct comprehensive optimization and systematic screening. For example:

Review the matching and compatibility of mobile and fixed equipment and analyze the operation conditions for passenger trains and over-line passenger trains to share common lines to realize the maximization of railway network resources; review whether the curve radius, grade and running speed meet topographic and geological conditions, traffic capacity and users' demands, and propose consultation opinions after technical and economic comparison and selection; review whether economical and intensive use of land as well as using less cultivated land is considered based on meeting requirements of traffic, production and safety protection; review the interfaces and anti-interference between bridges and roads, between tunnels and roads, between embankment and cutting, between civil works foundation and tracks, among four systems (including communication system, signaling system, power supply system and traction feeding system) and roads, bridges, tunnels and tracks as well as among four systems; review the linking and system integration of EMU and train control systems, pantograph–catenary relationship and wheel–rail relationship during design and construction, so as to propose consultation opinions on systematic optimization; review whether the overall functions, integrity, matching, compatibility and information sharing of systems (including civil works system, communication and signal system, traction feeding system, rolling stock (EMU) operation management system) meet requirements on system integration and systematic optimization, so as to optimize functional flow lines as per the principles of being practical, economic and artistic, to make the boarding and alighting of passengers become smooth, easy and separated, and to optimize section types and guarantee the safety when entering or departing from stations under the premise of a separation of passengers and trains (vehicles) as well as a convenient and ordered linking. See Fig. 3.49 for passenger flow organization at station.

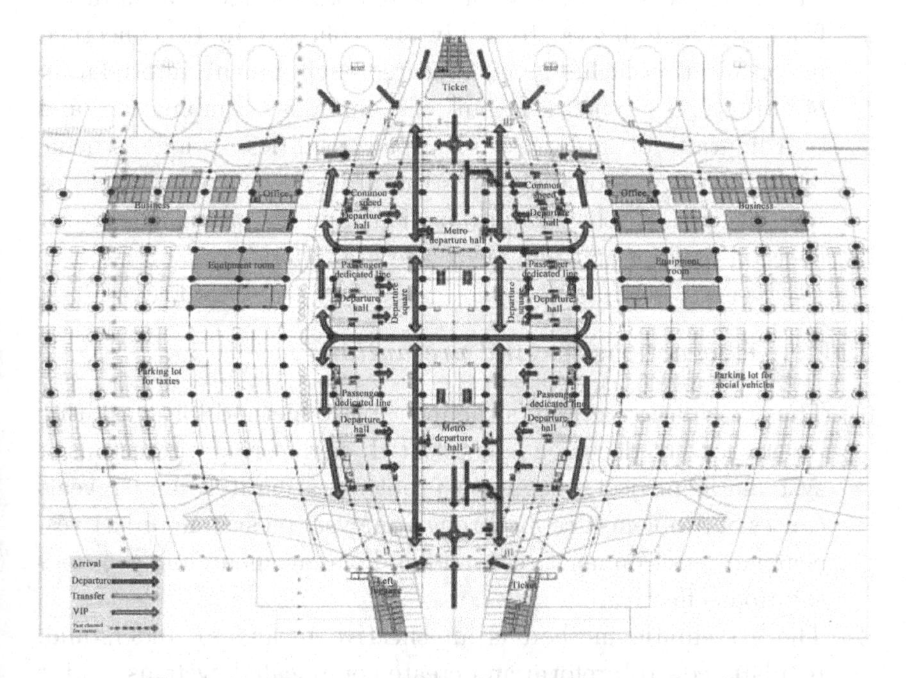

Fig. 3.49 Passenger Flow Organization at Station

(4) The modeling principle refers to making real systems into models whose forms and sizes meet people's demands and possibilities. The modeling principle is a guarantee for achieving optimization and a mean of the systematic method. For example:

Use various simulation software and visible platforms to collect engineering projects and environmental data, carry out data conversion based on CAD data, and show things that are hard to be seen by human eyes in the way of two dimensional or three dimensional forms through computer simulation software; provide multi-angle and three dimensional analysis platforms through software simulation and simulating visual angles and sights of people (such as sights of passengers on trains, sights from vehicles on roads along lines, sights of pedestrians along lines, sights from high-rise buildings along lines, sights of pedestrians on station platforms and sights of pedestrians in squares in front of stations); simulate railway subgrade, bridges, tunnels, stations and environment along lines to observe and analyze entire projects or engineering key points under consultation from different angles, to study the contents to be concerned in consultation including complicated environment, interrelation of structures, general layout of terminals, combination and coordination of railway and urban facilities, bridge scheme optimization, reasonability of structural schemes of buildings as well as station scheme optimization, and to provide bases for decision-making and judgment (see Fig. 3.50).

3.11.3 *Characteristics and precautions*

(1) The systematic method is the most basic method for consulting researches. It deems a complicated engineering project as a systematic project, diagnoses problems accurately and reveals causes of problems deeply through analyses on system objectives, elements, environment, resources and management, and proposes solutions effectively.
(2) The systematic method is an effective means to understand, regulate, control, reform and create complicated systems, and it

(a) Schematic Diagram for Sights of Passengers on Train

(b) Schematic Diagram for Sights from Vehicles on Roads

(c) Schematic Diagram for Sights from High-rise Buildings

(d) Schematic Diagram for Sights of Pedestrians on Station Platforms

(e) Schematic Diagram for Sights of Complicated Terminals from High-rise Buildings

(f) Schematic Diagram for Sights of Complicated Terminals from Pedestrians

Fig. 3.50 Three Dimensional Simulation Models

provides people with optimized schemes for system formulation to carry out optimized combination. The method breaks fixed forms of mechanical methods that only focus on analysis, and instruct people to think in a comprehensive way and explore new thoughts on technical development.

3.12 Feedback Method

The feedback method refers to using consultation opinions as well as the feedback opinions from the owner and the design institute on the consultation opinions for further consultation and reviewing their influences on projects especially on other associated disciplines and interfaces. For example:

The consulting agency collects, understands and seeks opinions of the owner and the design institute on the consultation, organizes evaluations occasionally, and proposes treatment opinions. Personnel responsible for the consultation of each discipline carry out rectification as per evaluation opinions, organize evaluations for non-conforming items, and propose treatment opinions as well as correction and precautionary measures. The consulting agency then carries out rectification as per evaluation opinions, organizes relevant organizations to evaluate non-conforming consultation conclusions identified during construction process, and proposes treatment opinions as well as correction and precautionary measures.

Upon completion of rectification, supplement and completion, non-conforming consultation results will be reviewed and signed level by level as per the review and signature procedures for consultation documents. The consulting agency uses information as the control basis and resource, regularly analyzes the conformity of consultation results through collecting and analyzing information from the owner and associated parties, and formulates corresponding improvement measures to continuously improve the engineering consultation quality.

3.13 Miscellaneous

In order to guarantee the railway engineering consultation quality and meet engineering construction requirements, the consulting agency should carry out consultation as per the above-mentioned consultation methods as well as pay attention to the following aspects:

(1) In terms of general control and organization measures, emphasizing the consultants' correct understanding and knowledge about the project significance and project characteristics under consultation.
(2) Selecting the personnel with experience in railway design and engineering consultation, especially the key technicians in the frontline, to participate in the consultation, clearly specifying the consultation and review personnel for the specific project, and setting up a technical adviser and expert group to take part in the research and decision-making on major technical issues.
(3) In terms of key technologies, making pertinent scientific research and monographic study and applying the research and study results to the engineering consultation.
(4) Taking advantage of the consulting agency's research, test, design and construction results of railways and the superiority of technical trend and guiding ideology to apply relevant results to the engineering consultation.
(5) Making systematic collection, analysis, comparison and study on relevant domestic and overseas technical standards to correctly know about the development tendency of railway technology.
(6) Enhancing technical exchanges with first-class design institutes and consulting agencies which have mature technologies and experience to know more about railway technologies and background, and conducting the consultation work in a creative way based on the actual conditions of the project under engineering consultation.
(7) Using advanced software for structural calculation and simulation analysis to strengthen the calculation and analysis and deepen the consultations on structural details and constructions.
(8) Laying emphasis on process consultation, check and investigation on the construction site by starting from the optimization of the plane and profile of the alignment, accurately expressing consultation comments, correctly knowing about technical

standards, and enhancing the review of safety and quality of works, quantities of works, design details, systems and interfaces.

To sum up, the consulting agency should conduct the consultation work with a scientific, harmonious and sustainable concept, in the principle of being scientifically rational, technically advanced and economically practical and in accordance with the requirements of building a conservation-minded society.

Chapter 4

Key Points of Railway Engineering Consultation

4.1 Design Optimization Consultation

Design optimization is carried out by the consulting agency to analyze the rationality of the construction standard, technical scheme and engineering measures applied, systematically optimize and perfect the design, give full play to the technological advantages of consultation, improve the railway construction quality, reflect the scientific development view implemented in railway construction in China, build a harmonious socialist society, adhere to the concept of a long-range program and quality first, and satisfy the general objective requirements for project construction quality by utilizing its technology and experience, making full use of and bringing into full play the experience and achievements in the design, research, consultation, construction, comprehensive test and operation of the existing railways, and also in combination with the project characteristics and the practical situation.

4.1.1 *Working basis*

(1) Relevant policies and regulations on infrastructure construction enacted by the State, the Ministry of Railways and the Ministry of Transport.

(2) Relevant technical standards, design specifications, acceptance standards, supervision regulations, construction specifications, regulations, environmental protection and water conservation requirements and completion acceptance measures on railway engineering construction formally enacted by the State and the Ministry of Railways.
(3) Relevant documents and replies on the project approved by the State, the Ministry of Railways and related departments, as well as relevant documents, notifications and meeting minutes on the project.
(4) *Preparation Method for Pre-Feasibility Study, Feasibility Study and Design Documents of Railway Construction Projects* (TB10504-2007).
(5) Reports of environment impact, water and soil conservation, geological hazard assessment, preservation of cultural relics, navigation, flood discharge capacity demonstration, etc. related to the project, as well as the approval opinions of the State and related departments.
(6) Related research results of railway.
(7) International advanced, mature, safe and reliable standard specifications.

4.1.2 *Consultation principle*

(1) Implement the policy of scientificity, harmony and sustainable development.
(2) Adopt international advanced, reliable, mature and applicable standards, specifications, theories and methods according to the main technical standards, social environment, geographical location, geological conditions and other characteristics of the project, and in compliance with the national laws, regulations, mandatory standards and the current industrial standards, specifications and regulations.
(3) Satisfy the general objective requirements of project construction quality and the requirements of constructing a conservation-minded society based on the comprehensive reference, digestion

and absorption of the mature experience and technical standards of railways in China and abroad.

(4) Realize the overall function and system integration of the project based on the principle of scientific rationality, safety and applicability, technical advancement, economic rationality, resource conservation, energy consumption reduction and environmental protection.

(5) Carry out systematic research, comprehensive analysis and repeated demonstration on the main technical standards, scheme, key parameters and engineering proposal of each discipline by utilizing various analysis methods including all kinds of simulation analysis.

(6) Lay emphasis on the optimization of engineering system design, technical standards, use function, overall design principle, discipline design principles and details, as well as discipline interface.

(7) Through review on the depth and breadth of survey and design, carry out comprehensive optimization of the general technical scheme of the whole line, design scheme of key and difficult construction works and critical works, major construction technology scheme and construction technology, as well as the standard drawing, general drawing and reference drawing that have a broad influence; focus on the optimization of plane and profile of the railway line, station site, proportions of ballastless track and bridge, track structure and four-electric system (communication, signal, traction feeding and electric power supply) design scheme.

4.1.3 *Overall requirements*

(1) Determine the key points of optimization design through the research and trends summary of design, construction and operation experience of Chinese high-speed railways and the six large scale speeding-up projects of Chinese railways, and in accordance with such characteristics of the project as the technical standards, line length, bridge and tunnel ratio, track type and laying lengths of different tracks, as well as distribution of soft subgrade; carry out further comparison, optimization

and improvement of relevant design according to the ideas of systematic design and integrated optimization.

(2) Based on the approved design of the previous stage and through the optimization design, ensure the engineering quality and safety, control the project investment, guarantee a reasonable construction period and lay emphasis on environmental protection and land saving.

(3) Give comprehensive consideration to the investment benefits and operation costs, optimize the layout and scale of stations and depots, save the investment and reduce the cost to maximize the benefits in combination with the railway transportation system reform and productivity distribution adjustment and according to the economy of cities along the railway line, passenger volume, railway transportation organization, carrying capacity and technical operation needs, as well as the engineering conditions and urban planning.

(4) Further improve the engineering safety and reliability, control the project investment and improve passenger comfort by optimizing the length of grade section on the profile of railway line, reasonably selecting the position and length of bridge and tunnel, avoiding high piers and large-span bridges, and avoiding frequent transition among bridge, tunnel, road and culvert; sort out the roads along the railway line in detail and for the crossing method and bridge type scheme adopted for the highway (road) and navigable river that have a great impact on the profile design elevation, verify the implementation status of relevant agreements, adjust the profile and optimize the design.

(5) Analyze the deepened results of the engineering geological work, bypass the unfavorable geological sections and focus on the research and optimization of the sections which can be bridge or road.

(6) For the sections with cities, towns and basic farmland, adjust and optimize the construction scheme of subgrade works, cut-fill adjustment and subgrade foundation treatment scheme, and determine a reasonable subgrade protection structure form according to the field investigation and verification conditions;

Fig. 4.1 A Long Span and Long Unit Continuous Girder Bridge Crossing a Navigable River

for the sections with high fill and deep excavation, make comparison between the subgrade and bridge scheme, and the subgrade and tunnel scheme and determine a reasonable structure form; optimize the engineering measures in karst area; optimize the subgrade foundation reinforcement measures and determine a reasonable post-construction settlement standard for the subgrade on the main line and station tracks inside the station.

(7) Optimize and unify the structure form and beam type of pier and abutment; optimize the pile foundation design through engineering pile test and other measures; determine the design and construction scheme of the bridges crossing rivers (as shown in Fig. 4.1) and roads according to local conditions.

(8) Strengthen the construction safety design of tunnel works and optimize the design of waterproofing and drainage in the tunnel inclined shaft, access adit and tunnel body.

(9) Optimize the drainage system design for subgrade, culvert and tunnel portal, especially for the highly-filled and deeply-excavated subgrade; pay attention to the disturbance and destruction to the original natural water system caused by railway construction, and prevent the infiltration of surface

water, rising of groundwater and water leakage from the lateral permeable water-proof interlayers; optimize the surface drainage slope, avoid the formation of new water gathering area and interception wall, etc., and optimize the setting range and length of waterproofing, drainage and interception facilities on the ground.

(10) Optimize the building scale, structure, plane and spatial layout, main dimensions, structural node, equipment selection and layout, pipeline diameter, pipeline layout, etc.

(11) Review the generality of design documents and the cohesive relationship among disciplines. By combining with the design documents of each discipline, inspect whether the selected construction schemes and interfaces among different disciplines are closely connected and focus on the systematic optimization of the interface and anti-interference between the bridge and subgrade, tunnel and subgrade, embankment and cutting, track foundation and track, four-electric system (communication, signal, traction feeding and electric power supply) and subgrade, bridge and tunnel as well as track, and among the four-electric systems, as well as of the relationships between EMU and train control system, pantograph and catenary, as well as wheel and rail.

(12) Optimize the construction organization scheme, reduce the scale of land use and the relocation quantity, and also reduce the triangular lands and marginal lands; optimize the process, construction period, construction and temporary transition measures and safety measures in construction organization design, as well as the construction technology and method, and schedule of each process.

(13) Protect the ecological environment, natural landscape and humanistic landscape and optimize the water and soil conservation as well as the measures of protection, disaster prevention and reduction, and pollution prevention and control for the ecologically sensitive districts by adhering to overall planning and satisfying the requirements of transport production and safety protection according to the requirements for implementation of

the national sustainable development strategy, intensive land use, occupation of less farmland and convenience for people's production and life; optimize the route selection and station site selection to bypass such environmentally sensitive areas as natural reserves, scenic spots, drinking water protected areas and key cultural relics sites under state protection; when passing cities or resident-inhabited areas, optimize the selection of speed target value and the noise and vibration reduction measures to satisfy the national environmental protection standards and requirements; optimize the protective measures combining the green plants on subgrade side slope with the engineering measures.

4.1.4 *General requirements*

(1) Review the implementation status of optimization carried out in accordance with relevant agreements, minutes, documents, reports, review and approval opinions.
(2) Optimize the discipline design principles and details.
(3) Review the implementation status of compulsory norms and standards of engineering construction and the rationality of its executive standards and specifications.
(4) Review whether the preparation contents, design depth and design quality of the design documents satisfy the requirements of *Preparation Method for Pre-feasibility Study, Feasibility Study and Design Documents of Railway Construction Projects* (TB10504-2007) and railway construction as well as the requirements specified in relevant specifications and regulations.
(5) Review the design documents for generality, systematicness and integrity and optimize the design scheme.
(6) Optimize the engineering measures, worksite technical measures and construction transition measures and strengthen the security measures.
(7) Strengthen and optimize the measures for fire prevention, savings, environmental protection, water conservation and preservation of cultural relics, and satisfy the requirements of relevant

policies, standards, specifications, regulations and for engineering construction.

(8) Review the work quantities and quantities of equipment, materials and land occupation for rationality, authenticity and reliability; check for the authenticity and reliability of relocation quantities and carry out comparative analysis with the work quantities of the previous stage.

4.1.5 *System integration*

(1) Requirements for system integration

System integration involves work related to overall design of system, interface technology, evaluation technology of safety, reliability and availability, electromagnetic compatibility evaluation technology and environmental condition technology, etc. The high-speed railway system consists of such subsystems as civil works, traction feeding, train operation control, high-speed train, operation dispatching and passenger service. It is featured by great investment, huge engineering scale, wide coverage, multi-discipline and domain, great technical difficulties, complex hierarchy and relation as well as great economic benefits and social benefits, thus becoming a typical super-large project. Therefore, at each stage (for example, design, construction and commissioning) of the construction process, attention should be paid to the standard matching and coordination, and the interface design coordination among systems as well as the matching and compatibility of fixed and mobile facilities according to the concept of system integration and by taking the system theory, control theory and synergetics theory as theoretical guidance, so as to realize optimization and creation of technology integration and satisfy the specific functional requirements of high-speed railways.

Review the elements of system integration, including the integrated cell, integrated mode, integrated interface, integrated condition and integrated environment of system integration, and evaluate the effects of system integration according to the basic theory of system integration. Especially the experience of the consulting agency in design, construction and management of system integration should be used as reference for the system integration of a project.

It is very important to develop basic principles and provide solutions for the problems that emerge at each stage of the project, and realize this by different ways in different levels and take appropriate countermeasures to reduce the integrated risk of the project. Such risk may occur in different forms during the entire project phase, for example:

- Poor design integration will lead to rework of design.
- Poor design integration will lead to rework and failure of construction.

For this purpose, the process and key points of system integration should be controlled according to the characteristics of high-speed railway system integration at different stages, so as to ensure the final effects of system integration can satisfy the demand. During the design consultation stage of system integration, focus should be placed on the analysis of demand for system integration and the refinement of the goal of system integration. During the construction and installation consultation stage of system integration, focus should be placed on the consultation for the division of subsystems, quality control management, technology and equipment method, and inspection and evaluation. Through system integration of high-speed railway, we should not only realize the single function of each independent system, but also achieve the functional integration among systems, as well as complete the overall function of the high-speed railway of carrying passengers in a safe, reliable, efficient and rapid manner, so as to maximize the service level for passengers.

The success of integration of a complex system often depends on the effective management of external and internal interfaces of system integration in the whole process of system integration by developing strict interface management procedures.

(2) Integration of four-electric systems (communication, signal, traction feeding and electric power supply)

For the high-speed railway, the system integration concept has been strengthened to ensure the realization of the overall construction goal. The project construction for such disciplines as communication, signaling, traction feeding and power supply of E&M works should be carried out based on the system integration mode.

The integration of four-electric systems refers to the functional integration and optimization of traction feeding system, power supply system, signaling system and communication system carried out in the stages of design, installation, commissioning and trial operation, with the aim of satisfying the functional requirements of the high-speed railway for the four-electric systems, as shown in Figs. 4.2 and 4.3. In Fig. 4.3, overall design of system refers to the overall design carried out for communication, signal, traction feeding and electric power supply, with the aim of system integration and optimization; interface technology refers to the unified management of the interfaces of various civil works and the interfaces among systems, and inside and outside the system; EMC evaluation technology refers to the systematic evaluation of the electromagnetic compatibility within a project; RAMS evaluation technology refers to the RAMS evaluation of the system, with the aim to improve the safety, reliability and availability of the system, where RAMS is developed to make such elements as reliability (R), availability (A), maintainability (M) and safety (S) run through the whole life cycle process from the envisioning phase to the end of life of the object system, and make them be implemented well as well as give consideration to the balance between them and economical efficiency; environmental condition technology refers to the comprehensive evaluation carried out for the environmental impacts of the system, with the aim to propose environmental control measures.

Through the four-electric system integration, the complicated external system interface can be converted into the internal system interface to facilitate the effective integration and functional integration of the system as well as the management of integration. Through system integration and

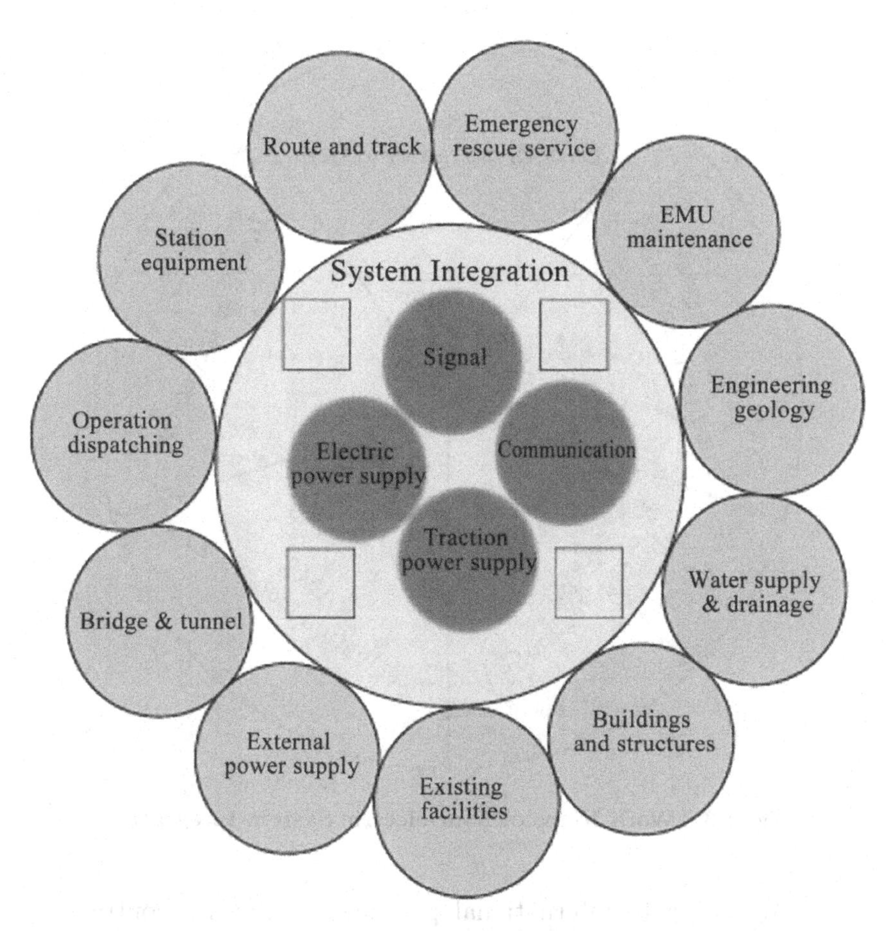

Fig. 4.2 Work Scope of Four-electric System Integration

through strengthening of process management, streamline management, reasonable organization, coordinated dispatching, optimal configuration and improvement of integrated level among four-electric systems in a project, various interface relationships are sorted out clearly and the civil works and E&M works are combined closely, so as to realize the seamless connection between design and construction, significantly reduce the altered design, take effective control of investment, improve the technical indexes, and fully realize and make more reasonable the system functions.

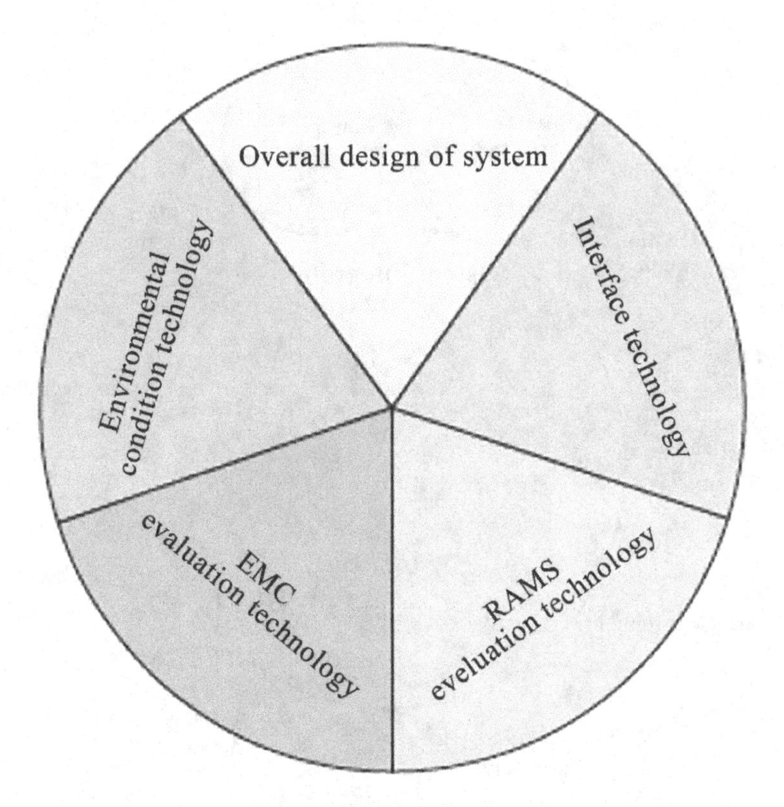

Fig. 4.3 Work Items of Four-electric System Integration

According to international practice, the general contracting mode should be adopted for system integration generally, and the system integrator is the core of technical support for the general contractor of system integration project and also the executor who aims at completing the turnkey project of design, supply and construction by introducing the system technology, integration technology, and design and construction technology for the employer. The system integrator shall take the overall responsibility for the technology, construction period, quality and safety of the communication, signal and traction feeding system works of high-speed railway, and shall be responsible for the following items: engineering design, equipment and material supply, construction and installation, commissioning (cooperation with the employer for integration test and trial operation), acceptance, technical service, training and technical service during the defects liability period; system scheme, system composition, system function and equipment performance; the matchability, compatibility and external interface of the supplied systems with the systems of EMU, subgrade, track, bridge, tunnel, station and yard, building, HVAC, disaster prevention and safety monitoring, integrated dispatching and information, and with the integrated systems of connection track, EMU depot and EMU servicing depot on other lines.

Key points of four-electric system integration:

(1) Carry out design optimization of the four-electric systems, especially the optimization of its interface functions, and find out the system function defect through definition, inspection and gap analysis of the internal and external interfaces of the four-electric system, so as to realize continuous improvement during the design, installation and commissioning stages.
(2) Implement effective management for the project during the project process, including engineering management, quality management, risk management, contract management, procurement management, document management, financial management and technical management.
(3) Develop strict test and inspection rules. Inspections include tests for all the equipment and systems of the four-electric systems as well as running tests related to train operation, so as to verify the realization of the objective of four-electric system integration.
(4) Realize Reliability, Availability, Maintainability and Safety (RAMS) index management during the whole life cycle of the project, ensure the RAMS requirements of the system have been taken into account during design of the four-electric systems, implement the RAMS requirements of the system during the project implementation and ensure the delivered system meets the RAMS requirements; find out and analyze all the unsafe factors that pose a threat to the safety of passengers, the public and operating management personnel,

and put forward the corresponding countermeasures to eliminate or alleviate these unsafe factors and minimize the risks to an acceptable level.

The RAMS analysis during the design stage mainly includes the following:

- Risk identification.
- Safety analysis.
- Put forward the countermeasures and minimize the risks to an acceptable level.
- Create a risk log which is used to record all the safety-related events and data and record related events and data during the whole life cycle of the project.
- Carry out design review to ensure all the risks associated with design are reduced to a level required by the employer.

The RAMS analysis during the project implementation stage mainly includes the following:

- Establish a safety tracking and evaluation system to carry out evaluation for the effects of all the design changes on the system safety in a timely manner.
- Set up a system safety reporting mechanism to record all the safety-related events and data that occurred during the construction stage to the risk log.
- Take all the feasible measures that can improve the system safety.

The RAMS analysis during the joint commissioning and trial operation stage mainly includes the following:

- Confirm the system safety requirements.
- Check the test results according to the safety requirements.
- Record all the safety-related events and data that occurred during the test stage to the risk log.
- Take all the feasible measures that can improve the system safety.
- Make a consultation for the above-mentioned key points of the four-electric system integration during the consultation process.

4.2 Construction Drawing Review Consultation

Construction drawing review is carried out by the consulting agency to focus on the review of the integrity and accuracy of design documents, the rationality and safety of major technical schemes, the safety and reliability of engineering measures, as well as the systematicness of connection between the discipline and works, by utilizing its technology and experience especially in the railway design and consultation, making full use of and bringing into full play the experience and achievements in the design, research, consultation, construction, comprehensive test and operation of the existing railways, and also considering the project characteristics and the practical situation, through the process consultation method and based on the review of the technical construction standards, design principles, and plane and profile of the line by taking the engineering quality and safety guarantee as the core. It is aimed to reduce deviations, errors, omissions and conflicts in design, take effective control of the engineering quality and safety, provide technical support for engineering construction, ensure the objective of safety, quality, construction period and investment of railway construction is under control and achieve the overall construction objective.

4.2.1 *Working basis*

(1) Relevant policies and regulations on infrastructure construction enacted by the State, the Ministry of Railways and the Ministry of Transport.
(2) Relevant technical standards, design specifications, acceptance standards, supervision regulations, construction specifications, regulations, environmental protection and water conservation requirements and completion acceptance measures on railway engineering construction formally enacted by the State and the Ministry of Railways.
(3) Relevant documents and replies on the project approved by the State, the Ministry of Railways and related departments, as well as relevant documents, notifications and meeting minutes on the project.

(4) *Preparation Method for Pre-Feasibility Study, Feasibility Study and Design Documents of Railway Construction Projects* (TB10504-2007).
(5) Reports of environment impact, water and soil conservation, geological hazard assessment, preservation of cultural relics, navigation, flood discharge capacity demonstration, etc. related to the project, as well as the approval opinions of the State and related departments.
(6) Related research results of railway.
(7) International advanced, mature, safe and reliable standard specifications.

4.2.2 *Consultation principle*

(1) Implement the policy of scientificity, harmony and sustainable development.
(2) Adopt international advanced, reliable, mature and applicable standards, specifications, theories and methods according to the main technical standards, social environment, geographical location, geological conditions and other characteristics of the project, and in compliance with the national laws, regulations, mandatory standards and the current industrial standards, specifications and regulations.
(3) Satisfy the general objective requirements of project construction quality and the requirements of constructing a conservation-minded society based on the comprehensive reference, digestion and absorption of the mature experience and technical standards of railways in China and abroad.
(4) Realize the overall function and system integration of the project based on the principle of scientific rationality, safety and applicability, technical advancement, economic rationality, resource conservation, energy consumption reduction and environmental protection.
(5) Based on the approved preliminary design and through the construction drawing review, ensure the engineering quality and safety, control the project investment, guarantee the construction progress and enforceability as well as the requirements for

durability, maintainability, environmental protection, intensive land use and urban planning.

(6) Lay emphasis on the review of engineering system design, technical standards, use function, overall design principle, discipline design principles and details, as well as discipline interface.

(7) Review whether the contents and depth of construction drawing conform to the provisions of the design contract and the contents of the design documents of construction drawing as well as the bidding requirements for the construction drawing.

4.2.3 *Overall requirements*

(1) Lay emphasis on the review of engineering system design, technical standards, use function, overall design principle, discipline design principles and details, discipline interface, depth and breadth of survey and design, key and difficult construction works and critical works, major construction technology scheme and construction technology, as well as the standard drawing, general drawing and reference drawing that have a broad influence, through the research and trends summary of design, construction and operation experience of Chinese high-speed railways and the six large-scale speeding-up projects of Chinese railways, and in accordance with such characteristics of the project as the technical standards, line length, bridge and tunnel ratio, track type and laying lengths of different tracks, as well as distribution of soft subsoil.

(2) Based on the approved design of the previous stage and through the review, ensure the engineering quality and safety, control the project investment, guarantee a reasonable construction period and satisfy the requirements for environmental protection and land saving.

(3) Focus on strengthening the review of new technology, new materials and new equipment by relying on the scientific and technological progress.

(4) Give comprehensive consideration to the investment benefits and operation costs and review whether the layout and scale of stations and depots satisfy the requirements for

systematicness, economic efficiency and rationality in combination with the railway transportation system reform and productivity distribution adjustment and according to the economy of cities along the railway line, passenger volume, railway transportation organization, carrying capacity and technical operation needs, as well as the engineering conditions and urban planning.

(5) Review whether the discipline design documents conform to the requirements of relevant regulations and specifications and whether the appraisal comments are strictly implemented; sort out the contents related to safety and reliability in the discipline design and check whether there is any safety and quality design defect; review the mutually submitted data among disciplines and check for the correctness of the design input data and the safety and reliability of the design scheme and design parameters.

(6) Review the selection of the length of grade section and the locations and lengths of bridges and tunnels on the profile of the railway line, avoid high piers and large-span bridges, avoid frequent transition among bridge, tunnel, road and culvert, further improve the engineering safety and reliability, control the project investment and improve passenger comfort; sort out the roads along the railway line in detail and verify the implementation status of relevant agreements.

(7) Analyze the deepened results of the engineering geological work and review whether the unfavorable geological sections are bypassed.

(8) Review the construction scheme of subgrade works, cut-fill adjustment and subgrade foundation treatment scheme as well as the subgrade protection structure form for rationality according to the field investigation and verification conditions; for the sections with high fill and deep excavation, make comparison and analysis with the bridge and tunnel scheme; review the overall pipeline layout in stations and yards, subgrade cross-section area, subgrade fillings, subgrade earth-rock proportion, earth-rock transport distance and foundation base treatment

for correctness and rationality, and review whether the greening scheme satisfies the requirements.

(9) Review the structure form of pier and abutment, the beam type selection and pile foundation design, and review whether to determine the construction scheme of the bridges crossing rivers and roads according to local conditions; review the foundation pit excavation and cofferdam protection for their safety and also review the requirements of flood control and navigation for bridge crossings.

(10) Review the construction safety design of tunnel works, the design of waterproofing and drainage of the tunnel inclined shaft and access adit, and verify the tunnel engineering measures for the emergency warning system, especially the engineering measures for unfavorable geological conditions. A picture of a high-speed railway tunnel is shown in Fig. 4.4.

(11) Review the drainage system design for subgrade, culvert and tunnel portal, especially for the highly-filled and deeply-excavated subgrade; pay attention to the disturbance and

Fig. 4.4 High-speed Railway Tunnel

destruction to the original natural water system caused by railway construction, and prevent the infiltration of surface water, rising of groundwater and water leakage from the lateral permeable water-proof interlayers; avoid the formation of new water-gathering area and interception wall, etc. by properly increasing the surface drainage slope and review the setting range and length of waterproofing, drainage and interception facilities on the ground.

(12) Review the building scale and standard, plane and spatial layout, main dimensions, structural node, equipment selection and layout, pipeline diameter, pipeline layout, etc.

(13) Review the overall functions, generality, matchability, compatibility and information sharing of the designed systems of civil works, communication signals, traction feeding, rolling stock (EMU) and operation management, to satisfy the requirements for system integration and system optimization; review the generality of design documents and the cohesive relationship among disciplines, especially the interface and anti-interference between the bridge and subgrade, tunnel and subgrade, embankment and cutting, track foundation and track, four-electric system (communication, signal, traction feeding and electric power supply) and road, bridge and tunnel as well as track, and among the four- electric systems, as well as the connection and systematic integration between EMU and train control system, pantograph and catenary, as well as wheel and rail in design and construction.

(14) Review the grade and actual demand of traction feeding, transformer connection mode, rationality and uniformity of specification and model selection for pole and column, interconnection between communication signals and adjacent lines, production and market supply of the main communication signal equipment, and matching between communication signals and onboard equipment.

(15) Review whether the design of communication, information and disaster prevention is unreasonable and lacks consideration for various natural disasters and review the design of the dispatch

communication system, GSM-R system, emergency communication and integrated video monitoring system for its safety and the design of the discipline interface for its rationality.

(16) Review the integrated grounding technical scheme for its safety and reliability and focus on reviewing whether the access scope of integrated grounding, cross section selection of through ground wire, grounding electrode connection and equipotential connection schemes, integrated grounding connection scheme for subgrade-bridge and subgrade-tunnel transition sections, and the grounding terminal reservation conform to the requirements of grounding protection of each system; review the measures of physical isolation of through ground wire with cable in the cable trough and the design scheme of interface with civil works for their rationality and enforceability.

(17) Review the process, construction period, construction and temporary transition measures and safety measures in construction organization design, as well as the construction technology and method, and schedule of each process; review whether the source of concrete sand and stone, borrow area and spoil area, waste disposal area, beam fabrication yard and track-laying base satisfy the principle of combination of permanent works and temporary works, and review for the rationality of their locations and scale, design depth, ballast source, laying of turnouts, track adjustment arrangement, commercial concrete selection and self-manufactured concrete equipment, etc.

(18) Review the intensive land use, occupation of less farmland, protection of the ecological environment, natural landscape and humanistic landscape, water and soil conservation, as well as the measures of protection, disaster prevention and reduction, and pollution prevention and control for the ecologically sensitive districts by adhering to overall planning and satisfying the requirements of transport production and safety protection; review whether environmentally sensitive areas including natural reserves, scenic spots, drinking water protected areas and key cultural relics sites under state protection are bypassed during route and site selection; review whether suitable speed

values as well as measures against noise and vibration are adopted when passing through cities or residential areas, and whether national environmental protection standards and requirements are met; review whether protection measures by combining green plants with projects are adopted or subgrade and side slope to meet requirements on appearance, environmental protection and water protection; review the measures for fire prevention, energy conservation, landscape, and vegetation protection and restoration.

4.2.4 *General requirements*

(1) Review the implementation status of relevant agreements, minutes, documents, reports, review and approval opinions.
(2) Review the discipline design principles and details.
(3) Review the implementation status of compulsory norms and standards of engineering construction and the rationality of its executive standards and specifications.
(4) Review whether the preparation contents, design depth and design quality of the design documents satisfy the requirements of *Preparation Method for Pre-feasibility Study, Feasibility Study and Design Documents of Railway Construction Projects* (TB10504-2007) and railway construction as well as the requirements specified in relevant specifications and regulations.
(5) Review the design documents for generality, systematicness and integrity and put forward specific conclusive suggestions on whether the overall design scheme can satisfy the requirements of the designed target speed.
(6) Review whether the engineering measures, worksite technical measures, construction transition measures and safety measures in the design documents satisfy the requirements of relevant specifications, regulations and for engineering construction.
(7) Review whether the measures for fire prevention, saving, environmental protection, water conservation and preservation of cultural relics satisfy the requirements of relevant policies, standards, specifications, regulations and for engineering construction.

(8) Review the work quantities and quantities of equipment, materials and land occupation for rationality, authenticity and reliability; check for the authenticity and reliability of relocation quantities and carry out comparative analysis with the work quantities of the previous stage.

(9) Review the selection of the standard design drawing, general design drawing and reference drawing for its rationality and conformity.

(10) Review for the generality of design documents and the interface design.

(11) Lay emphasis on process consultation, carry out intermediate inspection of the design process and put forward the consulting suggestions on supplementation, modification and completion for the implementation of design principles (details) and the special problems encountered in the design process.

4.2.5 *Interface management*

(1) Interface is classified into internal project interface, inter-project interface and external project interface (Fig. 4.5). Interface classification and determination shall be recorded well.

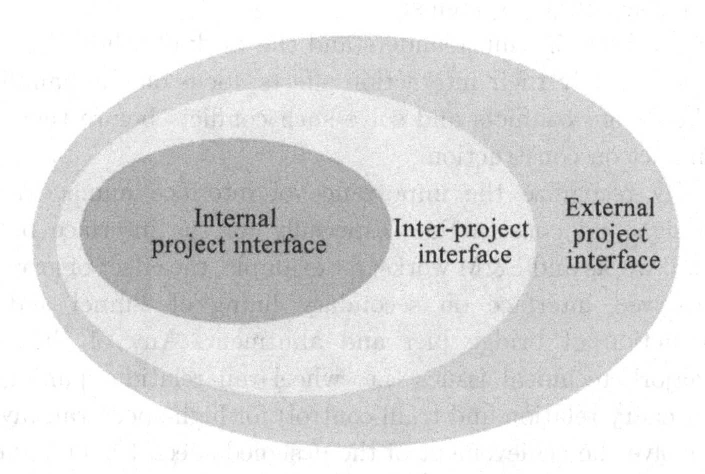

Fig. 4.5 Project Interface Relationship

(2) Carry out effective review and adjustment for the interface between individual works to allow for managing the interfaces between the stages, projects and management sections in the process of railway engineering construction.
(3) Review the interface among subgrade, track, power supply, OCS, signal, rolling stock and communication and establish a comprehensive technical support system; review the system engineering technology of the core systems of rolling stock, OCS and signal, as well as the project testing and commissioning technology.
(4) Review and adjust the design and management interface service.
(5) Hold the design and management interface meeting on a regular basis.
(6) Review in advance the interface between design and construction and between the interfacility design and construction.
(7) Set up a project interface management plan, prepare the project interface management procedures, establish a communication system, and review and adjust the design and management interface.
(8) Review and control the design interface.
(9) Review in advance the commissioning and trial operation interface among systems.
(10) Each discipline must understand the work of other disciplines, know clearly their interaction effects, focus on the handling of the design conflicts and solve such conflicts before they cause impact on construction.
(11) Fully recognize the importance of interface management in design and construction, especially for the interface between civil works and E&M works; for example, the effect of grounding reserved interface on secondary lining of tunnel and construction of bridge pier and abutment. Any of the "three major" technical issues (i.e. wheel–rail relation, pantograph-catenary relation and train control) for high-speed railways can involve the achievement of the designed speed target value and also some interface problems, such as bridge and ballastless

track, subgrade construction and OCS foundation construction, E&M works and integrated grounding, length of ballastless track and signaling and train control transfer distance, all of which should be given great attention in the consultation process.

4.3 Construction Process Consultation

Construction process consultation is carried out by the consulting agency to optimize the construction scheme, improve the engineering measures, identify the engineering hidden danger, improve the engineering quality and control the project investment by utilizing its technology and experience especially in railway design and consultation, making full use of and bringing into full play the experience and achievements in the design, research, consultation, construction, comprehensive test and operation of the existing railways, and also in combination with the actual site conditions. It is aimed to strengthen the control of engineering quality, safety, progress and investment, provide practical and effective guarantee to ensure the project is under control, to identify in a timely manner the main risks in project construction, establish a risk reporting and monitoring mechanism, eliminate or mitigate the construction and operation risks, ensure the objective of safety, quality, construction period and investment of railway construction is under control, and achieve the overall construction objective.

4.3.1 *Working basis*

(1) Relevant policies and regulations on infrastructure construction enacted by the State, the Ministry of Railways and the Ministry of Transport.

(2) Relevant technical standards, design specifications, acceptance standards, supervision regulations, construction specifications, regulations, environmental protection and water conservation requirements and completion acceptance measures on railway engineering construction formally enacted by the State and the Ministry of Railways.

(3) Relevant documents and replies on the project approved by the State, the Ministry of Railways and related departments, as well as relevant documents, notifications and meeting minutes on the project.
(4) *Preparation Method for Pre-Feasibility Study, Feasibility Study and Design Documents of Railway Construction Projects* (TB10504-2007).
(5) Reports of environment impact, water and soil conservation, geological hazard assessment, preservation of cultural relics, navigation, flood discharge capacity demonstration, etc. related to the project, as well as the approval opinions of the State and related departments.
(6) Related results of research on railways.
(7) International advanced, mature, safe and reliable standard specifications.

4.3.2 *Consultation principle*

(1) Implement the policy of scientificity, harmony and sustainable development.
(2) Adopt international advanced, reliable, mature and applicable standards, specifications, theories and methods according to the main technical standards, social environment, geographical location, geological conditions and other characteristics of the project, and in compliance with the national laws, regulations, mandatory standards and the current industrial standards, specifications and regulations.
(3) Satisfy the general objective requirements of project construction quality and the requirements of constructing a conservation-minded society based on the comprehensive reference, digestion and absorption of the mature experience and technical standards of railways in China and abroad.
(4) Realize the overall function and system integration of the project based on the principle of scientific rationality, safety and applicability, technical advancement, economic rationality, resource conservation, energy consumption reduction and environmental protection.

(5) Based on the field investigation and verification, review whether the design drawing and construction scheme comply with the site and surrounding construction conditions and whether the design scheme and engineering measures are reasonable and enforceable.
(6) Focus on the verification for the performance of the remaining and new engineering exploration works in combination with the implementation status of supplementary exploration in project implementation and the guidance of advanced prediction for design change.
(7) Focus on the discrimination and elimination of hidden dangers in engineering as well as the verification for the rationality and effectiveness of the waterproof and drainage system.
(8) Focus on the review of the necessity and rationality for the adjustment of implementation scheme as well as the investment changes caused by such adjustment.

4.3.3 *Overall requirements*

(1) Based on the field verification, exploration and annotation, review the key points and difficulties of the project, critical works, major construction technology scheme, construction technology, large-scale temporary works, etc.
(2) Review the engineering scheme and measures for their safety, reliability and rationality, the temporary protective structure for its safety and reliability, the engineering design scheme and scale for its enforceability and rationality, and the large-scale temporary works for its necessity, rationality and engineering implementation conditions. A railway construction site is shown in Fig. 4.6.
(3) Review whether the source of concrete sand and stone, borrow area and spoil area, waste disposal area, beam fabrication yard and track-laying and girder-erection base satisfy the principle of combination of permanent works and temporary works.
(4) Review the calculation of construction drawing investment and propose the measures for cause analysis and treatment of quantity difference.

Fig. 4.6 Railway Construction Site

4.3.4 *General requirements*

(1) Review whether the engineering measures, worksite technical measures, construction transition measures and safety measures satisfy the requirements of relevant specifications and regulations for engineering construction.
(2) Lay emphasis on process consultation and put forward the consulting suggestions on supplementation, modification and completion for the performance of design and the special problems encountered in the implementation process.
(3) Review the inspection results of engineering test items in the early stage of project implementation and put forward comments on their comprehensive popularization and application.
(4) Review the causes, responsibilities and investment changes of the altered design and make comments on the safety, reliability, rationality, technical and economic comparison of the altered design scheme or measures, as well as other engineering changes caused by such altered design.

(5) Review the measures for ensuring safety, quality, construction period, environmental protection and investment control.
(6) Review the major technical scheme, the site construction scheme for key works and the process, safety and quality assurance measures during construction.
(7) Review the restoration scheme for quality problems caused by design in the defects liability period.

4.3.5 *Interface management*

(1) Prepare the construction interface table based on the design interface table, including the technical interface among disciplines, technology and management interface among different construction contractors, technology and management interface among different design institutes, link-up of construction period and responsibility interface division.
(2) Establish a set of clear system interface documents and interface management plans to facilitate the interface management during construction, and make records of interface construction and commissioning.
(3) Hold the interface work meeting on a regular basis and strengthen the interface management.
(4) Review the system integration and interface management for its correctness, set the reliability, availability, maintainability and safety (RAMS) objectives of system integration and interface in the system and subsystems and define the reliability characteristics and the allowable downtime of the system as well as the time interval between system failures.
(5) Review whether the type selection of equipment is determined according to the set objective, whether the quantity meets the demand of the objective and whether the objective is included in the contracts signed with the suppliers to ensure the suppliers can provide the most reliable components and the performance of such components installed in the subsystems can achieve all the objectives of RAMS.

(6) Review the interfaces between bridge and subgrade, subgrade and track, four-electric system and subgrade, bridge and track, and among the four-electric systems for their connection in construction, and provide guiding opinions.

4.4 Special Consultation

Special consultation is carried out by the consulting agency to further absorb the advanced design concepts, standards and methods in China and abroad, take effective control of special key technologies, make a full comparison of the technical schemes, ensure the safety, reliability, economic efficiency and integrity of the technical schemes, provide technical support for engineering construction, ensure the objective of safety, quality, construction period and investment of railway construction is under control and achieve the overall construction objective through special consultation for the major technical issues affecting railway construction standards, the technical schemes involving structural safety, usability and economic rationality, as well as the major technical issues related to critical works or special structures, by utilizing its technology and experience especially in the railway design and consultation, making full use of and bringing into full play the experience and achievements in the design, research, consultation, construction, comprehensive test and operation of the existing railways.

4.4.1 *Working basis*

(1) Relevant policies and regulations on infrastructure construction enacted by the State, the Ministry of Railways and the Ministry of Transport.
(2) Relevant technical standards, design specifications, acceptance standards, supervision regulations, construction specifications, regulations, environmental protection and water conservation requirements and completion acceptance measures on railway engineering construction formally enacted by the State and the Ministry of Railways.

(3) Relevant documents and replies on the project approved by the State, the Ministry of Railways and related departments, as well as relevant documents, notifications and meeting minutes on the project.
(4) *Preparation Method for Pre-Feasibility Study, Feasibility Study and Design Documents of Railway Construction Projects* (TB10504-2007).
(5) Reports of environment impact, water and soil conservation, geological hazard assessment, preservation of cultural relics, navigation, flood discharge capacity demonstration, etc. related to the project, as well as the approval opinions of the State and related departments.
(6) Related research results of railway.
(7) International advanced, mature, safe and reliable standard specifications.

4.4.2 *Consultation principle*

(1) Implement the policy of scientificity, harmony and sustainable development.
(2) Adopt international advanced, reliable, mature and applicable standards, specifications, theories and methods according to the main technical standards, social environment, geographical location, geological conditions and other characteristics of the project, and in compliance with the national laws, regulations, mandatory standards and the current industrial standards, specifications and regulations.
(3) Satisfy the general objective requirements of project construction quality and the requirements of constructing a conservation-minded society based on the comprehensive reference, digestion and absorption of the mature experience and technical standards of railways in China and abroad.
(4) Realize the overall function and system integration of the project based on the principle of scientific rationality, safety and applicability, technical advancement, economic rationality, resource conservation, energy consumption reduction and environmental protection.

(5) Based on the preliminary design approval, ensure the engineering quality and safety, control the project investment, guarantee the construction progress and ensure environmental protection and land saving.
(6) Carry out systematic research, comprehensive analysis and repeated demonstration on the technical schemes, key parameters, structural details and engineering measures by utilizing various analysis methods including all kinds of simulation analysis.
(7) Focus on strengthening the review of new technology, new materials, new equipment and new process by relying on the scientific and technological progress; focus on the functions, generality, matchability, compatibility and information sharing of the system to satisfy the requirements for system integration and system optimization.
(8) Carry out consultation in combination with the design parameters of EMU, track structure and related parameters, special structural form, structural dynamic, static and deformation requirements, construction scheme, adjacent structural forms and construction methods, etc.

4.4.3 *Overall requirements*

(1) Carry out a profound and comprehensive consultation in aspects of the structural safety, reliability, generality, durability, adaptability, maintenance, handling of details, construction scheme, influence of related works, etc. by using the achievements, engineering experience and technology in the design, consultation, research and construction of railways in China and abroad, especially the structural works in special engineering environment, laying of ballastless track in special structures, special subgrade, special large-span bridges, and long tunnels.
(2) Carry out comprehensive and profound analysis and summarization with a focus on the following aspects by adopting special calculation software and methods:

Curve radius, transition curve alignment and length, distance between centers of tracks, lengths of intermediate straight line and circular curve, gradient difference between adjacent grade sections, length of grade section and dynamic characteristics of vertical curve and combination of horizontal and vertical curves.

Transition sections between different structures such as subgrade–bridge, bridge–bridge, bridge–tunnel and subgrade–culvert. Structural dynamic characteristics, deformation and settlement of typical structures, as well as stress of critical parts of special structures.

Safety, high smoothness, high comfort and economic rationality, construction conditions and maintenance, structural construction scheme and measures for design of special structures.

4.4.4 *General requirements*

(1) Review the implementation status of compulsory norms and standards of engineering construction and the rationality of its executive standards and specifications.
(2) Review whether the preparation contents, design depth and design quality of the design documents satisfy the requirements of *Preparation Method for Pre-feasibility Study, Feasibility Study and Design Documents of Railway Construction Projects* (TB10504-2007) and passenger dedicated line construction as well as the requirements specified in relevant specifications and regulations.
(3) Review the design documents for generality, systematicness and integrity and put forward specific conclusive suggestions on whether the overall design scheme can satisfy the requirements of the designed target speed.
(4) Review whether the design principles, parameter selection, mechanical model, integral structure, local structure and detail treatment are reasonable, whether the structural safety, reliability, applicability, enforceability and durability satisfy the requirements of relevant technical standards and whether the overall design is economical and reasonable.
(5) Review whether the engineering measures, worksite technical measures, construction transition measures and safety measures in the design documents satisfy the requirements of relevant specifications, regulations and for engineering construction.

(6) Review for the generality of design documents and the interface design.

4.4.5 *Key points of general special-subject consultation*

(1) Analyze and demonstrate the rationality of noise reduction design scheme and engineering measures for subgrade and bridge, and evaluate the effects of noise reduction design.
(2) According to the track form adopted, carry out rail–bridge interaction analysis and calculation for those large-span bridges or special structural bridges such as bridges with frequent span changes, tied-arch bridges and continuous girder arch bridges that are used for crossing the urban roads, and put forward suggestions or improvement opinions on the design optimization of track structure.
(3) Carry out analysis and calculation for the design and the additional longitudinal forces of continuously welded rail track within the turnout zone, demonstrate the safety, reliability and economic rationality of track structure design and put forward suggestions or improvement opinions on optimization.
(4) Based on the simulation analysis and research and the comprehensive evaluation for the dynamic performance of the transition sections among track, subgrade and bridge, and between subgrade and bridge, subgrade and culvert, subgrade and tunnel, subgrade and turnout, bridge and tunnel, bridges, tunnels, ballast track and ballastless track, propose the deformation distribution control standards, static and dynamic stability requirements, reliability requirements for long-term use, construction order and influence of actual settlement of the ballastless track, so as to ensure the safety and comfort of train operation on the whole high-speed railway.
(5) Study the influence of the settlement of structures such as subgrade, bridge, tunnel and culvert and the design and settlement of transition sections between embankment and cutting

and between ballast track and ballastless track on the laying of ballastless track.

(6) Carry out analysis and evaluation for the design of the typical transition sections with a large difference in strength and rigidity between subgrade and bridge, subgrade and culvert, subgrade and tunnel, embankment and cutting, semi-filled and semi-excavated subgrade, ballast track and ballastless track.

(7) Carry out research and evaluation for the technology of composite subgrade treatment for high-speed railways by use of CFG pile and its application conditions.

(8) Carry out analysis and evaluation for the economic rationality of the cut-fill adjustment principles having a great influence on construction, the improvement of unqualified filling materials and the long-distance transportation of qualified filling materials.

(9) Analyze the dynamic interaction between high-speed trains and track structure through the establishment of a dynamic analysis model of ballastless track on rolling stock-subgrade, and evaluate the dynamic performance of ballastless track on subgrade through simulation.

(10) Carry out safety risk analysis and evaluation for the design of urban bridges and long bridges, including the design of emergency passenger evacuation passageway and the light interference at parallel sections between high-speed railways and urban roads as well as high-speed highways.

(11) Based on the analysis for the influence of lateral rigidity of piers on the dynamic response of bridge, determine the relationship between the lateral rigidity of piers and the dynamic response of vehicle–bridge and propose a reasonable limit on the lateral rigidity of piers that satisfies the requirements for safe operation of high-speed trains and passenger riding comfort.

(12) Based on the structural form of bridges with turnouts at elevated stations for high-speed railways, put forward opinions to ensure the design scheme satisfies the requirements for safety, economic efficiency, rationality and beauty, suggestions

on improvement of bridge design scheme as well as design parameters and standards of bridges.

(13) Through analysis on the influence of deformation of bridges with turnouts on high-speed railways on the turnout stability and the influence of dynamic actions when the train passes through the turnout zone on a bridge at a high speed, evaluate the stability of the high-speed continuously welded turnouts on the elevated stations and bridges and the dynamic response of vehicle-turnout-bridge.

(14) Propose design optimization suggestions based on the dynamic performance analysis and analysis of deck form, track form and stress of structural critical parts, as well as analysis of safety and rationality of structural details, construction methods and measures for the special structural bridges.

(15) Carry out analysis and evaluation for the engineering measures and setting standards for remitting aerodynamics effects of tunnels, setting principles of fire rescue system, waterproof and drainage principles, treatment principles and construction scheme of tunnels in special geological environment (karst, goaf, gas, radioaction), setting principles of transition sections between bridge and tunnel and between tunnel and subgrade, as well as treatment principles of interfaces with other related disciplines.

(16) Carry out analysis and evaluation for the EMU utilization and maintenance management system and the distribution, scale, equipment and personnel allocation scheme of servicing facilities and maintenance equipment at EMU depots by combining with the operation management experience in China and abroad; carry out analysis and evaluation for the type selection, technological design and interface of the main equipment such as the wheelset diagnosis system, Computer Numerical Control (CNC) under-floor wheel lathe, bogie (wheelset) replacement device, inspection and testing equipment of each onboard control system, automated warehousing equipment and vacuum sewage discharge system.

(17) Carry out comprehensive evaluation for the overall technical scheme of integrated grounding. Carry out analysis and

evaluation for the schemes of grounding electrode setting, equipotential connection and grounding terminal setting for integrated grounding at special worksites; carry out analysis and evaluation on whether the access scope of integrated grounding system satisfies the requirements for grounding protection; carry out analysis and evaluation for the safety, rationality and enforceability of the interface design scheme for civil works.

(18) Analyze whether the structure of the communication network can satisfy the needs of development of such integrated services as voice, data and image, and whether the network design can satisfy the requirements for safety, reliability, flexibility and high efficiency; according to the operation organization structure and the requirements of each discipline, analyze whether the communication network can satisfy the requirements for safety, stability, reliability and flexibility of the train control, integrated dispatching and information system, and can meet the needs of development of such integrated services as voice, data and image, and propose an optimization scheme.

(19) Carry out evaluation for the settings of the disaster prevention and safety monitoring systems such as wind monitoring, rainfall monitoring and earthquake monitoring systems, train protection switches and telephones, access control system, foreign matter intrusion and unlawful entry protection system, fire monitoring and rail temperature monitoring systems that comprise the central system, station-level systems along the line and field monitoring system that are based on the high-speed railway communication transmission network.

(20) Carry out comprehensive evaluation for the traffic control signal system, train control system, interlock system, monitoring system, signal power supply system, lightning protection grounding and interface design; analyze and evaluate whether the type selection of each signal subsystem meets the operation needs; carry out analysis and evaluation for the safety, stability, reliability and maintainability of the technical scheme; carry out analysis and evaluation for adaptability with the format of the

signal systems such as traffic control system and train control system on adjacent lines.

(21) Carry out analysis and evaluation for the snow-melting devices of turnouts in cold regions and their power supply schemes.

(22) Carry out analysis and evaluation for concrete construction technology, anti-freezing and cracking prevention measures in cold regions.

4.4.6 *Key points of special consultation for track dynamics of railway line*

(1) Whole-process dynamic simulation analysis of plane and profile of railway line

(i) Carry out dynamic simulation calculation for the train–track coupling in the plane and profile design of railway line during the construction drawing design, with considerations for the parameters such as the designed plane curve radius, transition curve length, circular curve length, superelevation, intermediate straight line length, profile gradient, gradient difference between adjacent grade sections, vertical curve radius and grade section length in different sections of the whole line, and also for the combination of such parameters.

(ii) Carry out dynamic simulation calculation for the train–track coupling which takes the crosswind action into account. See Fig. 4.7.

(iii) Carry out dynamic simulation calculation for the train–track–bridge coupling on long simply-supported beam bridges with equal spans.

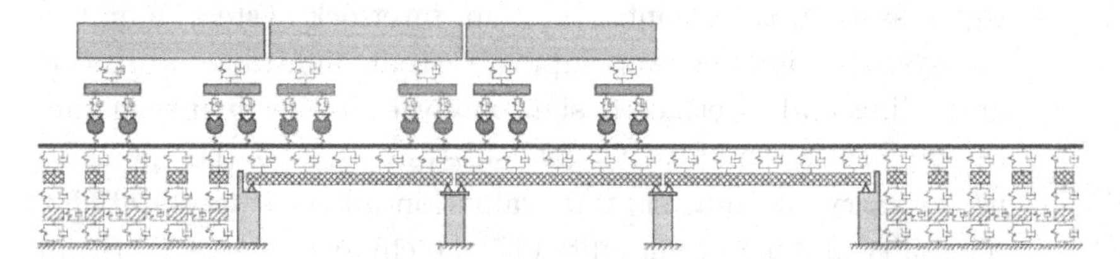

Fig. 4.7 Schematic Diagram of Train–Track Spatial Coupling Dynamic Model

(iv) Carry out analysis and evaluation for the wheel–rail vertical force, wheel–rail lateral force, increase of dynamic track gauge, wheel–rail wear index, equivalent taper, derailment coefficient, wheel unloading rate, vertical acceleration of car body, lateral acceleration of car body, vertical stability index of car body and lateral stability index of car body.
(v) Carry out evaluation for the safety, comfort and wheel–rail wear index of high-speed train operation.
(vi) Carry out evaluation for the plane and profile design parameters of railway line and the rationality and adaptability of their combination.
(vii) Carry out evaluation for the relationships between different superelevations, centrifugal force, vehicle vibration and the critical wind speed, and also for the safety of train operation on curves at a high speed.
(viii) For the long simply-supported beam bridges with equal spans, carry out evaluation for the deflection, natural vibration frequency, lateral amplitude, vibration acceleration, displacement of structures, influence of the ratio of span and vehicle length, motion mass and damping effect.

(2) Dynamic simulation analysis for transition sections (see Fig. 4.8)
(i) Dynamic simulation analysis for train–track coupling at typical embankment-cutting transition section.
(ii) Dynamic simulation analysis for train–track–bridge coupling at typical subgrade-bridge transition section.
(iii) Dynamic simulation analysis for train–track coupling at typical subgrade-tunnel transition section.
(iv) Dynamic simulation analysis for train–track coupling at typical transition section of subgrade turnout zone.
(v) Dynamic simulation analysis for train–track–bridge coupling at typical transition section of bridge turnout zone.
(vi) Dynamic simulation analysis for train–track coupling at typical ballast track-ballastless track transition section.
(vii) Dynamic simulation analysis for train–track coupling at typical subgrade-culvert transition section.

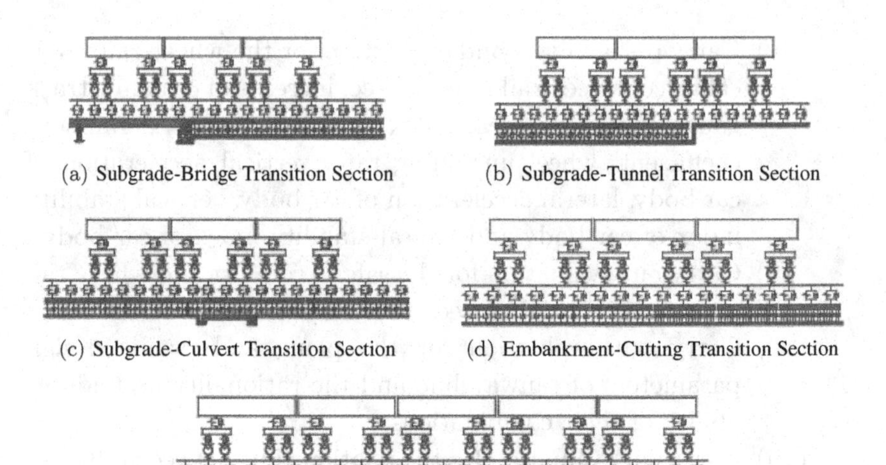

(e) Transition Section of Short Subgrade between Bridge and Tunnel

Fig. 4.8 Dynamic Model of Transition Sections for High-speed Trains

(viii) Dynamic simulation analysis for train–track–bridge coupling at typical transition section of short subgrade between bridges.

(ix) Dynamic simulation analysis for train–track–bridge coupling at typical transition section of short subgrade between bridge and tunnel.

(x) Dynamic simulation analysis for train–track coupling at typical transition section of short subgrade between tunnels.

(xi) Carry out evaluation for the safety and comfort of high-speed train operation as well as the rationality and adaptability of design of transition sections through analysis of such data as the wheel–rail vertical force, wheel–rail lateral force, derailment coefficient, wheel unloading rate, vertical acceleration of car body, lateral acceleration of car body, Sperling stability index, vertical deflection of bridge, lateral amplitude of bridge, vertical acceleration of bridge, lateral acceleration of bridge, lateral stability coefficient of bridge, rail support pressure, dynamic stress of CA mortar, dynamic stress at the top of backfilled concrete, dynamic stress on subgrade surface and rail deflection change rate.

4.4.7 *Key points of special consultation for special structural bridge*

(1) Key points of design consultation for cable-stayed bridges with steel-trussed girders (see Figs. 4.9 and 4.10)

(i) Carry out analysis and evaluation for the rationality of general layout of bridge according to the engineering environmental conditions, requirement of clearance under bridge, flood discharge requirements, overall

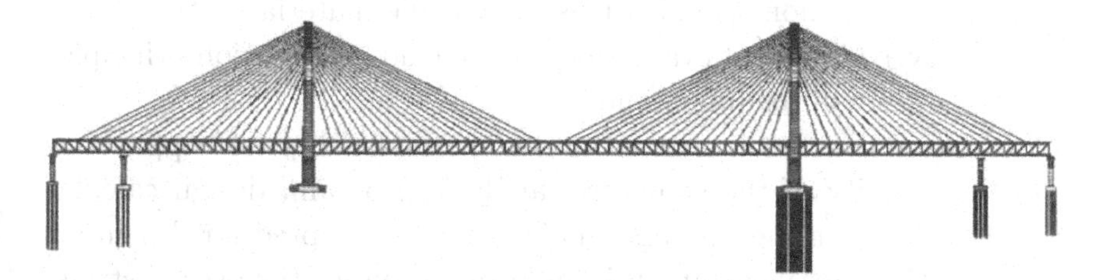

Fig. 4.9 Full-bridge Calculation Model of Cable-stayed Bridge with Steel-trussed Girders

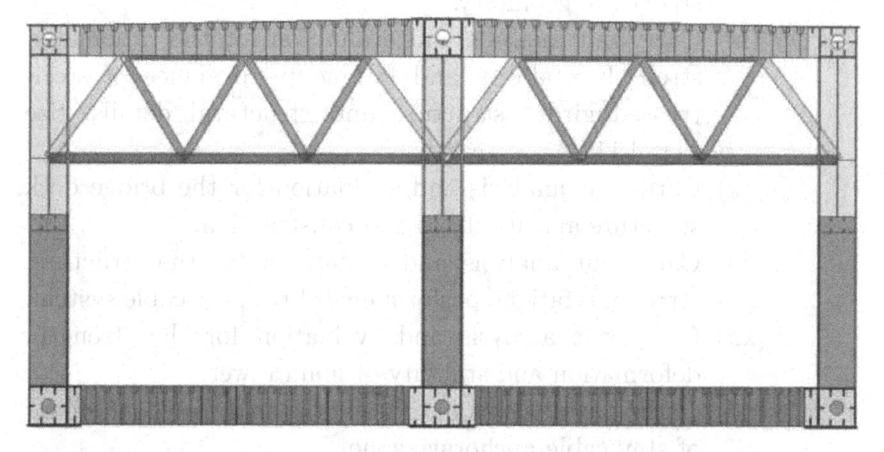

Fig. 4.10 Cross Section of Main Girder of Cable-stayed Bridge with Steel-trussed Girders

landscape of bridge, construction conditions and line conditions.

(ii) Carry out analysis and evaluation for the rationality of deck layout form, running safety, passenger comfort and structural durability.

(iii) Carry out analysis and evaluation for the rationality of the structural forms of pier, girder, tower and stay cable.

(iv) Carry out evaluation for the rationality of selection of steel strength, weldability, heat treatment and chemical components, etc.

(v) Carry out evaluation for the rationality of selection of concrete grade, steel bar specifications, prestressed tendon specifications, stay cable materials, etc.

(vi) Carry out evaluation for the load distribution principles and loading standards.

(vii) Carry out evaluation for the rationality and applicability of the structural analysis program, design calculation method and analysis model adopted for design.

(viii) Carry out analysis and evaluation for the overall stress, deformation and stability of bridge with considerations for the nonlinear effects of structural materials and structural geometry.

(ix) Carry out analysis and evaluation for the rationality, strength, stability and fatigue performance of steel-trussed girder structure and structural details. See Fig. 4.11.

(x) Carry out analysis and evaluation for the bridge deck structure and its stress and construction.

(xi) Carry out analysis and evaluation for the structure, stress and fatigue performance of the stay cable system.

(xii) Carry out analysis and evaluation for the strength, deformation and stability of main tower.

(xiii) Carry out analysis and evaluation for the local effects of stay cable anchorage zone.

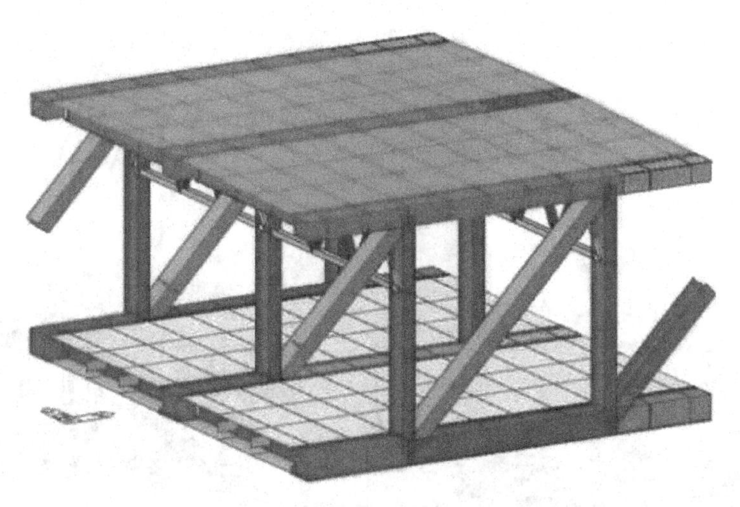

Fig. 4.11 Schematic Diagram of Truss Segment of Cable-stayed Bridge with Steel-trussed Girders

(xiv) Carry out analysis and evaluation for the connection structure between tower and stay cable and between stay cable and girder, the stress and its impact on the fatigue performance of the structure. See Fig. 4.12.

(xv) Carry out analysis and evaluation for the stress, settlement and deformation of substructure and foundation.

(xvi) Carry out evaluation for the rationality, safety and reliability and maintainability of technical standards, design parameters, structural form and stress of the damping device.

(xvii) Carry out evaluation for measures for bridge protection and collision prevention as well as safety measures for navigation and driving under bridge.

(xviii) Carry out evaluation for the rationality of the hydraulic damping device determined through analysis of earthquake, temperature change, braking force and other longitudinal forces.

(xix) Carry out analysis and evaluation for the structural dynamic characteristics.

(xx) Carry out evaluation for the rationality, applicability and validity of calculation results of the train types and

Fig. 4.12 Calculation Model for Connection between Tower and Girder

parameters, track irregularity spectrum, train–bridge–track boundary conditions, etc. that are adopted in the analysis and calculation of dynamic response of train–bridge coupling.

(xxi) Carry out analysis and evaluation for the wind-induced vibration and rain-induced vibration during the bridge completion phase and construction phase.

(xxii) Carry out evaluation for the design of ship impact loads and collision prevention during the bridge completion phase and construction phase.

(xxiii) Carry out analysis and evaluation for the dynamic response of train–bridge coupling with consideration for the impact of temperature change on geometry of bridge deck and the wind action.

(xxiv) Carry out analysis and evaluation for the seismic performance during the bridge completion phase and construction phase.

(xv) Carry out evaluation for the structural stress, deformation, settlement, dynamic performance and their impacts on structures such as the track.

(xxvi) Carry out evaluation for the effect of longitudinal force of continuously welded rail tracks on bridge and the setting principles, position and form of expansion joint, as well as their impacts on train operation.

(xxvii) Carry out evaluation for the anti-corrosion measures of steels.

(xxviii) Carry out evaluation for the design of concrete structure durability.

(xxix) Carry out evaluation for anti-corrosion and waterproof design for bridge deck.

(xxx) Carry out evaluation for the accessibility, inspectability, maintainability, serviceability and replaceability of bridge structure.

(xxxi) Carry out evaluation for the engineering measures for reduction of impact of operation noise and wind on trains and for stay cable protection in case of train derailment.

(xxxii) Carry out health monitoring and evaluation.

(xxxiii) Carry out evaluation for the fabrication, erection sequence and process design of steel girder.

(xxxiv) Carry out evaluation for the design of stay cable tensioning sequence, bridge alignment control and stress adjustment.

(xxxv) Carry out evaluation for the construction scheme and construction organization design of foundation, tower and girder.

(xxxvi) Carry out evaluation for the rationality, safety, economic efficiency and applicability of foundation construction scheme and technical measures, especially the large-scale deep-water cofferdam and marine equipment, safety guarantee measures and pile forming technology for complex geology.

(xxxvii) Carry out evaluation for the rationality of layering for mass concrete pouring and the measures for reduction of hydration heat in early stage and control of temperature rise and temperature difference; carry out

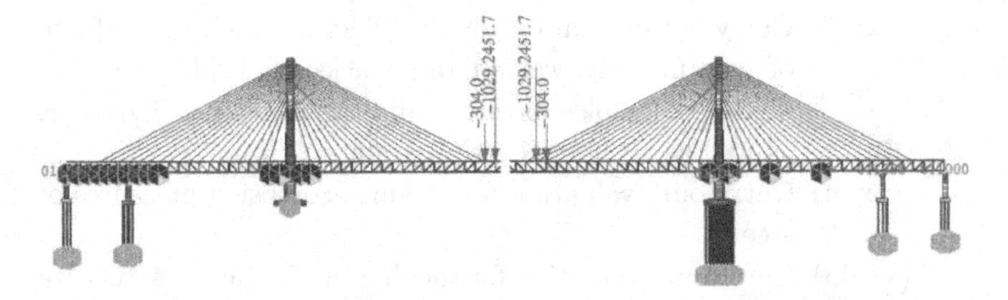

Fig. 4.13 Stress Analysis at the Largest Single-cantilever State

Fig. 4.14 Arch Bridge with Steel-trussed Girders

stress analysis and evaluation with consideration for the actual constraints and concrete performance (Fig. 4.13).

(2) Key points of design consultation for arch bridges with steel-trussed girders (see Fig. 4.14)

(i) Carry out analysis and evaluation for the rationality of general layout of bridge according to the engineering environmental conditions, requirement of clearance under bridge, flood discharge requirements, overall landscape of bridge, construction conditions and line conditions.

(ii) Carry out analysis and evaluation for the rationality of deck layout form, running safety, passenger comfort and structural durability.

(iii) Carry out analysis and evaluation for the rationality of the structural forms of pier, girder, arch and suspender.

(iv) Carry out evaluation for the rationality of selection of steel strength, weldability, heat treatment and chemical components, etc.

(v) Carry out evaluation for the rationality of selection of concrete grade and steel bar specifications, etc.

(vi) Carry out evaluation for the load distribution principles and loading standards.

(vii) Carry out evaluation for the rationality and applicability of the structural analysis program, design calculation method and analysis model adopted for design.

(viii) Carry out analysis and evaluation for the overall stress, deformation and stability of girder and arch.

(ix) Carry out analysis and evaluation for the rationality, strength, stability and fatigue performance of steel-trussed girder structure and structural details.

(x) Carry out analysis and evaluation for the bridge deck structure and its stress and construction.

(xi) Carry out analysis and evaluation for the arch structure, stress and fatigue performance.

(xii) Carry out analysis and evaluation for the suspender structure, stress and fatigue performance.

(xiii) Carry out analysis and evaluation for the connection structure between arch and girder, and between suspender and arch and girder, the stress and its impact on the fatigue performance of the structure.

(xiv) Carry out analysis and evaluation for the local effects of connection between suspender and arch and girder.

(xv) Carry out analysis and evaluation for strength and rigidity of piers.

(xvi) Carry out analysis and evaluation for foundation stress, settlement and deformation.

(xvii) Carry out evaluation for measures for bridge protection and collision prevention as well as safety measures for navigation and driving under bridge.

(xviii) Carry out analysis and evaluation for the structural dynamic characteristics.

(xix) Carry out evaluation for the rationality, applicability and calculation results validity of the train types and parameters, track irregularity spectrum, train–bridge–track boundary conditions, etc. that are adopted in the analysis and calculation of dynamic response of train–bridge coupling.

(xx) Carry out evaluation for the design of ship impact loads and collision prevention during the bridge completion phase and construction phase.

(xxi) Carry out analysis and evaluation for the dynamic response of train–bridge coupling with consideration for the impact of temperature change on geometry of bridge deck and wind action.

(xxii) Carry out analysis and evaluation for the seismic performance during the bridge completion phase and construction phase.

(xxiii) Carry out evaluation for structural stress, deformation, settlement, dynamic performance and their impacts on structures such as the track.

(xxiv) Carry out evaluation for the effect of longitudinal force of continuously welded rail tracks on bridge and the setting principles, position and form of expansion joint, as well as their impacts on train operation.

(xxv) Carry out evaluation for the anti-corrosion measures of steels.

(xxvi) Carry out evaluation for the design of concrete structure durability.

(xxvii) Carry out evaluation for anti-corrosion and waterproof design for bridge deck.

(xxviii) Carry out evaluation for the accessibility, inspectability, maintainability, serviceability and replaceability of bridge structure.

(xxix) Carry out evaluation for the engineering measures for reduction of impacts of operation noise and wind on trains and for suspender protection in case of train derailment.

(xxx) Carry out health monitoring and evaluation.

(xxxi) Carry out evaluation for the construction scheme and construction organization design of foundation, pier, girder and arch.

(xxxii) Carry out evaluation for the fabrication, erection sequence and process design of steel girder, arch and suspender.

(xxxiii) Carry out evaluation for bridge alignment control.

(xxxiv) Carry out evaluation for the rationality, safety, economic efficiency and applicability of foundation construction scheme and technical measures, especially the large-scale deep-water cofferdam and marine equipment, safety guarantee measures and pile forming technology for complex geology.

(xxxv) Carry out evaluation for the rationality of layering for mass concrete pouring and the measures for reduction of hydration heat in early stage and control of temperature rise and temperature difference; carry out stress analysis and evaluation with consideration for the actual constraints and concrete performance.

(3) Key points of design consultation for steel-tied arch bridges (see Fig. 4.15)

(i) Carry out analysis and evaluation for the rationality of general layout of bridge according to the engineering environmental conditions, requirement of clearance under bridge, flood discharge requirements, overall landscape of bridge, construction conditions and line conditions.

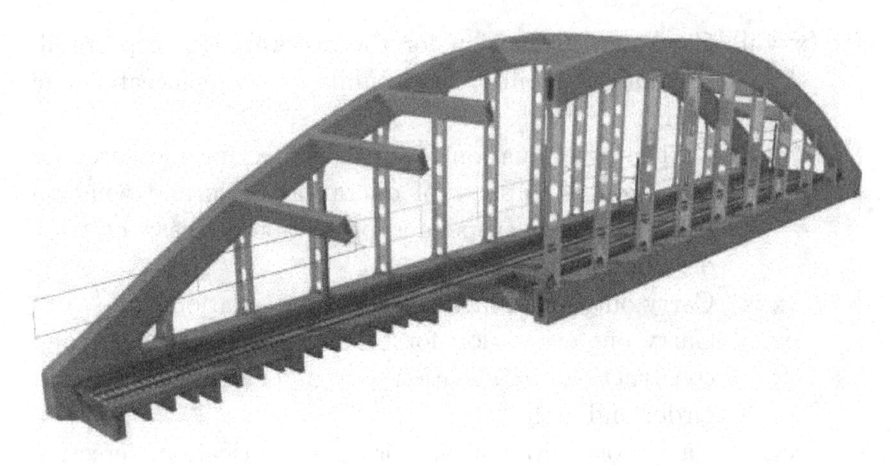

Fig. 4.15 Structural Diagram of Through Type Steel-tied Arch Bridge

(ii) Carry out analysis and evaluation for the rationality of deck layout form, running safety, passenger comfort and structural durability.

(iii) Carry out analysis and evaluation for the rationality of the structural forms of pier, girder, arch and suspender.

(iv) Carry out evaluation for the rationality of selection of steel strength, weldability, heat treatment and chemical components, etc.

(v) Carry out evaluation for the rationality of selection of concrete grade and steel bar specifications, etc.

(vi) Carry out evaluation for the load distribution principles and loading standards.

(vii) Carry out evaluation for the rationality and applicability of the structural analysis program, design calculation method and analysis model adopted for design.

(viii) Carry out analysis and evaluation for the overall stress, deformation and stability of girder and arch.

(ix) Carry out analysis and evaluation for the rationality, strength, stability and fatigue performance of steel box girder structure and structural details.

(x) Carry out analysis and evaluation for the bridge deck structure and its stress and construction.

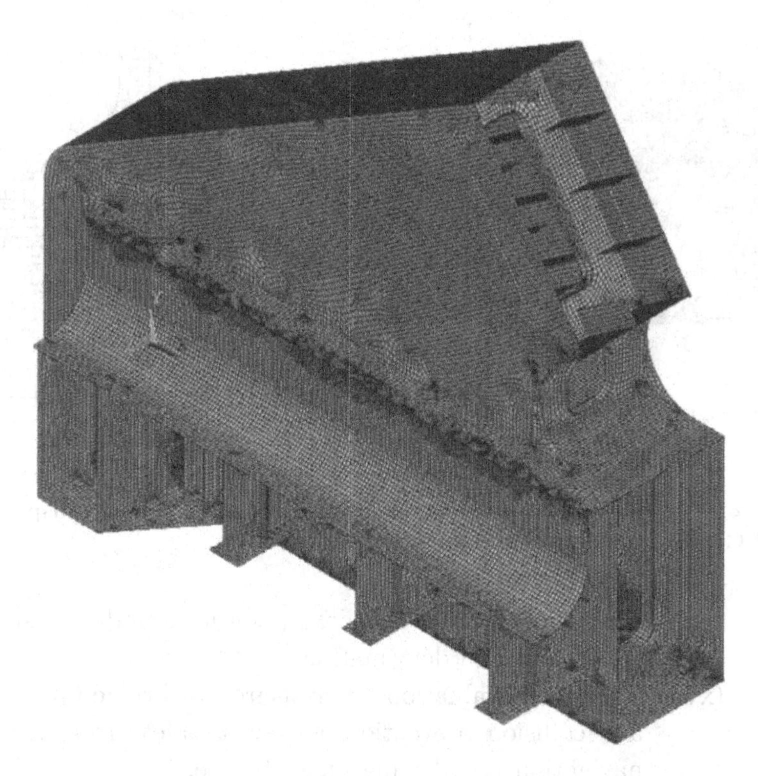

Fig. 4.16 Analysis Unit Division of Arch Springing

(xi) Carry out analysis and evaluation for the arch structure, stress and fatigue performance.

(xii) Carry out analysis and evaluation for the suspender structure, stress and fatigue performance.

(xiii) Carry out analysis and evaluation for the connection structure between arch and girder and between suspender and arch and girder, the stress and its impact on the fatigue performance of the structure. The analysis unit division of arch springing is shown in Fig. 4.16.

(xiv) Carry out analysis and evaluation for the local effects of connections between suspender and arch and between suspender and girder (see Fig. 4.17).

(xv) Carry out analysis and evaluation for strength and rigidity of piers.

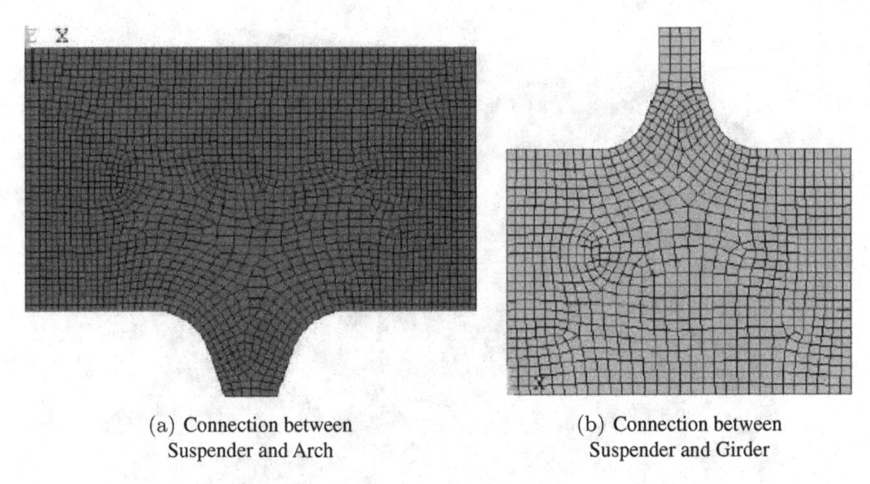

(a) Connection between Suspender and Arch

(b) Connection between Suspender and Girder

Fig. 4.17 Analysis Unit Division of Connections between Suspender and Girder and between Suspender and Arch

(xvi) Carry out analysis and evaluation for foundation stress, settlement and deformation.

(xvii) Carry out evaluation for measures for bridge protection and collision prevention as well as safety measures for navigation and driving under bridge.

(xviii) Carry out analysis and evaluation for the structural dynamic characteristics.

(xix) Carry out evaluation for the rationality, applicability and calculation results validity of the train types and parameters, track irregularity spectrum, train–bridge–track boundary conditions, etc. that are adopted in the analysis and calculation of dynamic response of train–bridge coupling.

(xx) Carry out evaluation for the design of ship impact loads and collision prevention during the bridge completion phase and construction phase.

(xxi) Carry out analysis and evaluation for the dynamic response of train–bridge coupling with consideration for the impact of temperature change on geometry of bridge deck and wind action.

(xxii) Carry out analysis and evaluation for the seismic performance during the bridge completion phase and construction phase.

(xxiii) Carry out evaluation for the structural stress, deformation, settlement, dynamic performance and their impacts on structures such as the track.

(xxiv) Carry out evaluation for the effect of longitudinal force of continuously welded rail tracks on bridge and the setting principles, position and form of expansion joint, as well as their impacts on train operation.

(xxv) Carry out evaluation for the anti-corrosion measures of steels.

(xxvi) Carry out evaluation for the design of concrete structure durability.

(xxvii) Carry out evaluation for anti-corrosion and waterproof design for bridge deck.

(xxviii) Carry out evaluation for the accessibility, inspectability, maintainability, serviceability and replaceability of bridge structure.

(xxix) Carry out evaluation for the engineering measures for reduction of impacts of operation noise and wind on trains and for suspender protection in case of train derailment.

(xxx) Carry out health monitoring and evaluation.

(xxxi) Carry out evaluation for the construction scheme and construction organization design of foundation, pier, girder and arch.

(xxxii) Carry out evaluation for the fabrication, erection sequence and process design of steel girder, arch and suspender. Figure 4.18 shows the lifting of arch segments of through type steel-tied arch bridge.

(xxxiii) Carry out evaluation for bridge alignment control.

(xxxiv) Carry out evaluation for the rationality, safety, economic efficiency and applicability of foundation construction scheme and technical measures, especially the large-scale deep-water cofferdam and marine equipment, safety

Fig. 4.18 Lifting of Arch Segments of Through Type Steel-tied Arch Bridge

guarantee measures and pile forming technology for complex geology.

(xxxv) Carry out evaluation for the rationality of layering for mass concrete pouring and the measures for reduction of hydration heat in early stage and control of temperature rise and temperature difference; carry out stress analysis and evaluation with consideration for the actual constraints and concrete performance.

(4) Key points of design consultation for prestressed concrete continuous girder arch bridges (Fig. 4.19)

(i) Carry out analysis and evaluation for the rationality of general layout of bridge according to the engineering environmental conditions, requirement of clearance

Fig. 4.19 Effect Picture of Prestressed Concrete Continuous Girder Arch Bridge

under bridge, flood discharge requirements, overall landscape of bridge, construction conditions and line conditions.

(ii) Carry out analysis and evaluation for the rationality of deck layout form, running safety, passenger comfort and structural durability.

(iii) Carry out analysis and evaluation for the rationality of the structural forms of pier, girder, arch and suspender.

(iv) Carry out evaluation for the rationality of selection of steel strength, weldability, heat treatment and chemical components, etc.

(v) Carry out evaluation for the rationality of selection of concrete grade, steel bar specifications, prestressed tendon specifications, etc.

(vi) Carry out evaluation for the load distribution principles and loading standards.

(vii) Carry out evaluation for the rationality and applicability of the structural analysis program, design calculation method and analysis model adopted for design.

(viii) Carry out analysis and evaluation for the overall stress, deformation and stability of girder and arch.

(ix) Carry out analysis and evaluation for the rationality of girder structure and the effects of its strength, creep deformation, temperature deformation and differential settlement on structural stress and deformation.

(x) Carry out analysis and evaluation for the arch structure, stress and deformation.

(xi) Carry out analysis and evaluation for the suspender structure, stress and fatigue performance.

(xii) Carry out analysis and evaluation for the connection structure between arch and girder and between suspender and arch and girder, and its stress.

(xiii) Carry out analysis and evaluation for the local effects of connection between suspender and arch and girder.

(xiv) Carry out evaluation for the control of initial tensioning force of suspender and shrinkage as well as creep of main girder.

(xv) Carry out analysis and evaluation for strength and rigidity of piers.

(xvi) Carry out analysis and evaluation for foundation stress, settlement and deformation.

(xvii) Carry out evaluation for measures for bridge protection and collision prevention as well as safety measures for navigation and driving under bridge.

(xviii) Carry out analysis and evaluation for the structural dynamic characteristics.

(xix) Carry out evaluation for the rationality, applicability and calculation results validity of the train types and parameters, track irregularity spectrum, train–bridge–track boundary conditions, etc. that are adopted in the analysis and calculation of dynamic response of train–bridge coupling.

(xx) Carry out evaluation for the design of ship impact loads and collision prevention during the bridge completion phase and construction phase.

(xxi) Carry out analysis and evaluation for the dynamic response of train–bridge coupling with consideration for the impact of temperature change on geometry of bridge deck and the wind action.

(xxii) Carry out analysis and evaluation for the seismic performance during the bridge completion phase and construction phase.

(xxiii) Carry out evaluation for structural stress, deformation, settlement, dynamic performance and their impacts on structures such as the track.

(xxiv) Carry out evaluation for the effect of longitudinal force of continuously welded rail tracks on bridge and the setting principles, position and form of expansion joint, as well as their impacts on train operation.

(xxv) Carry out evaluation for the anti-corrosion measures of steels.

(xxvi) Carry out evaluation for the design of concrete structure durability.

(xxvii) Carry out evaluation for waterproof design for bridge deck.

(xxviii) Carry out evaluation for the accessibility, inspectability, maintainability, serviceability and replaceability of bridge structure.

(xxix) Carry out evaluation for the engineering measures for reduction of impacts of wind on trains and for suspender protection in case of train derailment.

(xxx) Carry out health monitoring and evaluation.

(xxxi) Carry out evaluation for the construction scheme and construction organization design of foundation, pier, girder and arch.

(xxxii) Carry out evaluation for bridge alignment control.

(xxxiii) Carry out evaluation for the rationality, safety, economic efficiency and applicability of foundation construction scheme and technical measures, especially the large-scale deep-water cofferdam and marine equipment, safety guarantee measures and pile forming technology for complex geology.

Fig. 4.20 Effect Picture of a Continuous Girder Bridge with Prestressed Concrete Turnouts

(xxxiv) Carry out evaluation for the rationality of layering for mass concrete pouring and the measures for reduction of hydration heat in early stage and control of temperature rise and temperature difference; carry out stress analysis and evaluation with consideration for the actual constraints and concrete performance.

(5) Key points of design consultation for bridges with prestressed concrete turnouts (see Fig. 4.20)

(i) Carry out analysis and evaluation for the rationality of general layout of bridge according to the engineering environmental conditions, requirement of clearance under bridge, overall landscape of bridge, construction conditions and turnout layout conditions.

(ii) Carry out analysis and evaluation for the rationality of deck layout form, running safety, passenger comfort and structural durability.

(iii) Carry out analysis and evaluation for the rationality of the structural forms of pier and girder.

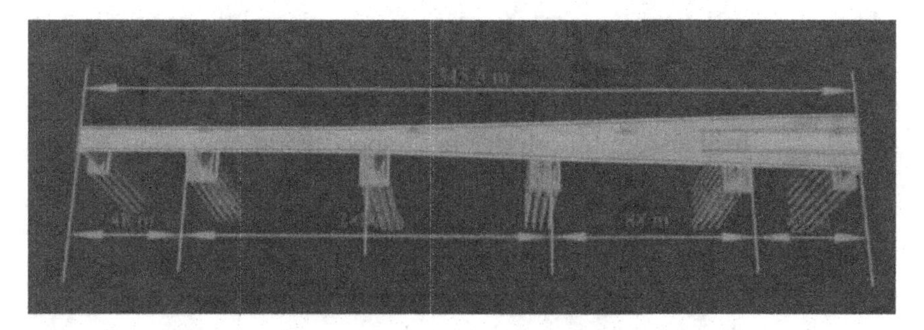

Fig. 4.21 Calculation Model for a Continuous Girder Bridge with Prestressed Concrete Turnouts

(iv) Carry out evaluation for the rationality of selection of concrete grade, steel bar specifications, prestressed tendon specifications, etc.

(v) Carry out evaluation for the load distribution principles and loading standards.

(vi) Carry out evaluation for the rationality and applicability of the structural analysis program, design calculation method and analysis model adopted for design.

(vii) Carry out analysis and evaluation for the rationality of girder structure and the effects of its strength, creep deformation, temperature deformation and differential settlement on structural stress and deformation, as well as the effects of bridge deformation on turnout stability (Fig. 4.21).

(viii) Carry out analysis and evaluation for strength and rigidity of piers.

(ix) Carry out analysis and evaluation for foundation stress, settlement and deformation.

(x) Carry out evaluation for measures for bridge protection and collision prevention as well as safety measures for navigation and driving under bridge.

(xi) Carry out analysis and evaluation for the structural dynamic characteristics.

(xii) Carry out evaluation for the rationality, applicability and validity of the calculation results of the train types and parameters, track parameters, train–bridge–track boundary conditions, etc. that are adopted in the analysis and calculation of dynamic response of train–turnout–bridge coupling under high-speed train operation.

(xiii) Carry out analysis and evaluation for the seismic performance.

(xiv) Carry out evaluation for structural stress, deformation, settlement, dynamic performance and their impacts on structures such as the track.

(xv) Carry out evaluation for the effect of longitudinal force of continuously welded rail tracks on bridge and its impact on train operation.

(xvi) Carry out evaluation for the design of concrete structure durability.

(xvii) Carry out evaluation for waterproof design for bridge deck.

(xviii) Carry out evaluation for the accessibility, inspectability, maintainability, serviceability and replaceability of bridge structure.

(xix) Carry out health monitoring and evaluation.

(xx) Carry out evaluation for the construction scheme and construction organization design of foundation, pier and girder.

(xxi) Carry out evaluation for bridge alignment control.

(xxii) Carry out evaluation for the rationality, safety, economic efficiency and applicability of foundation construction scheme and technical measures, especially the large-scale deep-water cofferdam and marine equipment, safety guarantee measures and pile forming technology for complex geology.

(xxiii) Carry out evaluation for the rationality of layering for mass concrete pouring and the measures for reduction of hydration heat in early stage and control of temperature rise and temperature difference; carry out stress analysis and evaluation with consideration for the actual constraints and concrete performance.

(6) Key points of consultation for construction organization and construction scheme of long-span prestressed concrete bridges

(i) Carry out analysis and evaluation for the rationality of general construction scheme of bridge according to the engineering environmental conditions, and the requirement of clearance under bridge, construction conditions, construction organization, construction technology, construction method, engineering measures, safety guarantee and risk prevention after considering the construction and protection measures.

(ii) Carry out analysis and evaluation for the rationality of the construction schemes of designed cast-in-place cantilever construction with cradles, cast-in-place support, hauling construction and swivel erection construction according to the overall construction organization design of the whole line, construction period of bridge, construction period risk evaluation, feasibility of structure and implementation, economic efficiency of construction scheme, safety of navigation and driving under bridge, large-scale temporary facilities, etc.

(iii) Carry out analysis and evaluation for the construction methods, #0 section construction, bracket or support section construction at the top of sidespan piers, closure section construction, construction procedure, beam body alignment control, construction transition measures, safety protection scheme and construction technology, etc.

(iv) Carry out evaluation for the design of hanging cradle, the design, check and calculation and test of #0 section and load-carrying structures such as bracket or support at the top of sidespan piers.

(v) Carry out evaluation for the design, installation and movement of hanging cradle, midspan and sidespan closure schemes, selection and preparation of durable concrete, continuous girder alignment control, prestressed duct reservation and grouting, tendon pushing and prestressed tensioning, girder surface leveling and construction of embedded parts.

(vi) Carry out evaluation for major working procedure and safety and quality accident prevention measures, emergency assistance plan, protective shed and hanging cradle protection at major parts, especially for the traveling of hanging cradle and the provision of double-layer safety protection facilities at the front end at the worksites with limited clearance conditions.

(vii) Make suggestions on the adjustment and optimization of design scheme, improvement of process and operation procedure, improvement of construction efficiency, optimization and combination of construction schemes as well as other aspects according to the site conditions.

(viii) Carry out evaluation for the detailed structure (such as the beam-end detailed structure that affects the prestressed construction and end seal concrete quality), continuous structure segment length, layout of reinforcements with difficulties in construction and installation as well as concrete pouring and tamping, prestressed pipe form and materials, closure section locking scheme and measures. Figure 4.22 shows the cantilever pouring construction of a long-span prestressed concrete bridge.

(7) Key points of design consultation for high-pier and long-span prestressed concrete bridges

(i) Carry out analysis and evaluation for the rationality of general layout and structural form of bridge according to the engineering environmental conditions, requirement of clearance under bridge, flood discharge requirements, overall landscape of bridge, construction conditions and line conditions.

(ii) Carry out evaluation for the rationality of deck layout form, running safety, passenger comfort and structural durability.

Fig. 4.22 Cantilever Pouring Construction of a Long-span Prestressed Concrete Bridge

(iii) Carry out analysis and evaluation for the rationality of the structural forms of pier and girder.

(iv) Carry out evaluation for the rationality of selection of concrete grade, steel bar specifications, prestressed tendon specifications, etc.

(v) Carry out evaluation for the rationality and applicability of the structural analysis program, design calculation method and analysis model adopted for design.

(vi) Carry out analysis and evaluation for the overall stress, deformation and stability of bridge.

(vii) Carry out analysis and evaluation for the connection structure between pier and girder and its stress effect.

(viii) Carry out analysis and evaluation for the stress, settlement and deformation of substructure and foundation.

(ix) Carry out evaluation for measures for bridge protection and collision prevention as well as safety measures for navigation and driving under bridge.

(x) Carry out analysis and evaluation for the structural dynamic characteristics.

(xi) Carry out evaluation for the rationality, applicability and validity of calculation results of the train types and parameters, track irregularity spectrum, train–bridge–track boundary conditions, etc. that are adopted in the analysis and calculation of dynamic response of train–bridge coupling.

(xii) Carry out evaluation for the design of ship impact loads and collision prevention during the bridge completion phase and construction phase.

(xiii) Carry out analysis and evaluation for the dynamic response of train–bridge coupling with consideration for the impact of temperature change on the geometry of bridge deck and wind action.

(xiv) Carry out evaluation for the high-pier temperature field distribution rules, temperature gradient change rules, temperature stress calculation and analysis methods, as well as the size, shape and spacing of air vents of high hollow piers.

(xv) Carry out analysis and evaluation for seismic performance during the bridge completion phase and construction phase and for the safety control methods of train operation on bridge under the action of common earthquake.

(xvi) Carry out evaluation for the design classification, principles and methods of seismic ductility, as well as seismic structural measures of piers.

(xvii) Carry out evaluation for structural stress, deformation, settlement and dynamic performance as well as their impacts on track and other structures, especially for the impacts of beam-end deformation, rotation angle and expansion on ballastless tracks at a great slope.

(xviii) Carry out evaluation for the effect of longitudinal force of continuously welded rail tracks on bridge and the setting principles, position and form of expansion joint, as well as their impacts on train operation.

(xix) Carry out evaluation for the design of concrete structure durability.

(xx) Carry out evaluation for anti-corrosion and waterproof design for bridge deck.

(xxi) Carry out evaluation for the accessibility, inspectability, maintainability, serviceability and replaceability of bridge structure.

(xxii) Carry out evaluation for the aerodynamic characteristics of bridge and vehicle under the action of strong wind on a high-pier bridge, wind speed threshold or train speed threshold for train operation on the bridge, as well as the engineering measures for reducing the impact of wind on trains.

(xxiii) Carry out health monitoring and evaluation.

(xxiv) Carry out evaluation for design of bridge alignment control.

(xxv) Carry out evaluation for the construction scheme and construction organization design of foundation and girder. A high-pier and long-span prestressed concrete bridge is shown in Fig. 4.23.

(xxvi) Carry out evaluation for the rationality, safety, economic efficiency and applicability of foundation construction scheme and technical measures, especially the large-scale deep-water cofferdam and marine equipment, safety guarantee measures and pile forming technology for complex geology.

(xxvii) Carry out evaluation for the rationality of layering for mass concrete pouring and the measures for reduction of hydration heat in early stage and control of temperature rise and temperature difference; carry out stress analysis and evaluation with consideration for the actual constraints and concrete performance.

Fig. 4.23 High-pier and Long-span Prestressed Concrete Bridge

(8) Key points of design consultation for long-span concrete arch bridges in mountainous area (see Figs. 4.24 and 4.25)

(i) Carry out analysis and evaluation for the rationality of general layout of bridge according to the engineering environmental conditions, requirement of clearance under bridge, flood discharge requirements, overall landscape of bridge, construction conditions and line conditions.

(ii) Carry out analysis and evaluation for the rationality of deck layout form, running safety, passenger comfort and structural durability.

(iii) Carry out analysis and evaluation for the structural forms of pier, girder and arch, and the rationality of their relationships.

(iv) Carry out evaluation for the rationality of selection of steel strength, weldability, heat treatment and chemical components, etc.

(v) Carry out evaluation for the rationality of selection of concrete grade, steel bar specifications, prestressed tendon specifications, etc.

Fig. 4.24 Effect Picture of a Long-span Concrete Arch Bridge

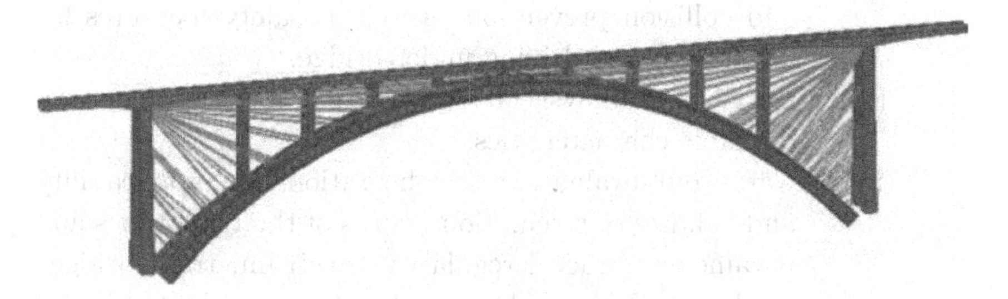

Fig. 4.25 Analysis Model for Large-span Concrete Arch Bridge

(vi) Carry out evaluation for the load distribution principles and loading standards.

(vii) Carry out evaluation for the rationality and applicability of the structural analysis program, design calculation method and analysis model adopted for design.

(viii) Carry out analysis and evaluation for the overall stress, deformation and stability.

(ix) Carry out analysis and evaluation for the rationality of arch structure and the effects of its strength, creep deformation and temperature deformation on structural stress and deformation.

(x) Carry out analysis and evaluation for the arch structure, stress and deformation and stability.

(xi) Carry out analysis and evaluation for the structure, stress, deformation, concrete creep, stability and wind resistance of arch and its frame during the arch construction.

(xii) Carry out analysis and evaluation for the connection structure between arch and pier and its stress.

(xiii) Carry out analysis and evaluation for the girder structure, stress and deformation.

(xiv) Carry out analysis and evaluation for strength and rigidity of piers.

(xv) Carry out analysis and evaluation for foundation stress and deformation.

(xvi) Carry out evaluation for measures for bridge protection and collision prevention as well as safety measures for navigation and driving under bridge.

(xvii) Carry out analysis and evaluation for the structural dynamic characteristics.

(xviii) Carry out evaluation for the rationality, applicability and validity of calculation results of the train types and parameters, track irregularity spectrum, train–bridge–track boundary conditions, etc. that are adopted in the analysis and calculation of dynamic response of train–bridge coupling.

(xix) Carry out evaluation for the design of ship impact loads and collision prevention during the bridge completion phase and construction phase.

(xx) Carry out analysis and evaluation for the dynamic response of train–bridge coupling with consideration for the impact of temperature change on geometry of bridge deck and wind action.

(xxi) Carry out analysis and evaluation for the seismic performance during the bridge completion phase and construction phase.

(xxii) Carry out evaluation for structural stress, deformation, settlement, dynamic performance and their impacts on structures such as the track.
(xxiii) Carry out evaluation for the effect of longitudinal force of continuously welded rail tracks on bridge and the setting principles, position and form of expansion joint, as well as their impacts on train operation.
(xxiv) Carry out analysis and evaluation for the laying conditions of ballastless tracks on bridge and the reasonable structural form of ballastless tracks.
(xxv) Carry out evaluation for the anti-corrosion measures of steels.
(xxvi) Carry out evaluation for the design of concrete structure durability.
(xxvii) Carry out evaluation for waterproof design for bridge deck.
(xxviii) Carry out evaluation for the accessibility, inspectability, maintainability, serviceability and replaceability of bridge structure.
(xxix) Carry out evaluation for the engineering measures for reduction of impacts of wind on trains and for suspender protection in case of train derailment.
(xxx) Carry out health monitoring and evaluation.
(xxxi) Carry out evaluation for the construction scheme and construction organization design of foundation, pier, girder and arch.
(xxxii) Carry out evaluation for bridge alignment control.
(xxxiii) Carry out evaluation for the rationality, safety, economic efficiency and applicability of foundation construction scheme and technical measures, especially the large-scale deep-water cofferdam and marine equipment, safety guarantee measures and pile forming technology for complex geology.
(xxxiv) Carry out evaluation for the rationality of layering for mass concrete pouring and the measures for reduction of hydration heat in early stage and control of temperature

rise and temperature difference; carry out stress analysis and evaluation with considerations for the actual constraints and concrete performance.

4.4.8 *Key points of special consultation for special tunnel*

(1) Key points of design consultation for river-crossing shield tunnel

(i) Carry out analysis and evaluation for the design principles, construction clearance, cross section and overall design according to the engineering environmental conditions, construction conditions, line conditions and technical standards as well as requirements for clear width, clear height, seismic resistance and float resistance of tunnel.

(ii) Carry out analysis and evaluation for the design principles, standards and codes, and design loads of hidden section, open section and working shaft structure, workbench materials and structural calculation, structure construction and construction measures, design contents and requirements of construction drawing.

(iii) Carry out design and calculation for the design principles, standards and codes, design load, engineering materials and lining structure of shield tunnel, design and construction of transverse passageway, design and construction of track structure in the tunnel, as well as design and evaluation of the shield tunnel.

(iv) Carry out evaluation for the waterproof design principles, standards and codes, as well as waterproof design of hidden section, working shaft and shield section.

(v) Carry out evaluation for the ventilation design principles, standards and codes, design parameters, ventilation mode, ventilation system running mode, ventilation system control, tunnel environment monitoring, sound insulation and vibration attenuation.

(vi) Carry out evaluation for water supply and drainage design principles, standards and codes, as well as water supply system, drainage system, fire fighting system, pipe material selection, water pump control mode and drainage equipment.

(vii) Carry out evaluation for the codes and standards, design scope, contents and principles adopted for lighting design, technical requirements for lighting power supply system, type selection and laying methods of cable and wire, technical specifications for type selection of electrical protection equipment, design contents and requirements of lighting construction drawing.

(viii) Carry out evaluation for the codes and standards, design principles and technical requirements adopted for durability design.

(ix) Carry out evaluation for the general requirements, composition and functional requirements as well as subsystems of the comprehensive disaster prevention system.

(x) Carry out evaluation for the general requirements of comprehensive monitoring design, system function and system scheme, division of central control part and interface, design contents and requirements of monitoring construction drawing.

(xi) Carry out analysis and evaluation for the stability design of tunnel structure, design of transverse passageway, durability design of tunnel shield segment, design of tunnel ventilation and rescue system and design of open excavation sections at tunnel entrance and exit.

(xii) Carry out evaluation for the construction quality and technical acceptance standards adopted.

(xiii) Carry out evaluation for construction scheme, risk management, shield supplier and construction organization design, especially for the schemes of shield tunneling machine design, segment prefabrication, departure and reception of shield, installation of shield tunneling machine, removal of shield tunneling machine, shield driving, cutter replacement of shield, shield seal, synchronous grouting of shield, transverse passageway construction, meeting of shield

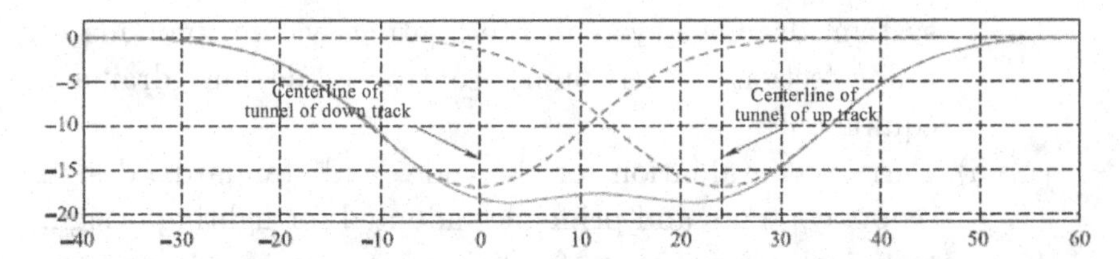

Fig. 4.26 Prediction Curve of Highway Settlement Caused by Shield Construction of a River-crossing Tunnel (Unit: mm)

tunneling machines in tunnel and construction survey, as well as for the construction safety measures for the tunnel works (Fig. 4.26).

(2) Key points of design consultation for multi-track station tunnel

(i) Carry out evaluation for the design reliability, practicability, maintainability and safety, as well as the technical standards and specifications adopted for tunnel design, excavation and support, ventilation and rescue.

(ii) Carry out evaluation for exploration of engineering geological and hydrogeological conditions, aboveground structures and underground pipeline distribution.

(iii) Carry out evaluation for the basic clearance requirement for maintenance, rescue, engineering technical operations and other purposes adopted in design.

(iv) Carry out analysis and evaluation for aerodynamics of high-speed train operation in the tunnel, cross section form and area of the tunnel.

(v) Through the thermal environment aerodynamics analysis, carry out evaluation for the changes of air temperature field, pressure field and velocity field of tunnel, chimney effects at different sections, temperature distribution at connection area, temperature distribution of station tracks, air pressure transition analysis, thermal reflow at tunnel entrance and ventilation effect, and optimize the ventilation shaft position and size, as well as the design of natural ventilation and mechanical ventilation.

(vi) Carry out evaluation for the effect of micro-pressure wave at the portal on the portal environment.

(vii) Carry out evaluation for the setting, air exchange standard and frequency of operation ventilation, maintenance ventilation and disaster prevention ventilation of tunnel, as well as for the control and emission of flue gas and pollutants.

(viii) Carry out evaluation for the functions and requirements of disaster-preventing and alarm (fire, water volume, smoke, accident) system, disaster prevention and rescue system, civil works, ventilation, firefighting, lighting, signal, communication and power supply, especially for the fire risk, fire and smoke control, induction and alarm system, fire extinguishing system setting, fire safety disposal, fire rescue passage and flue gas emission methods.

(ix) Carry out evaluation for firefighting, methods and time of escape, evacuation path, rescue, emergency lighting, evacuation channel setting and service gallery setting as well as measures for safe and quick personnel evacuation without lighting.

(x) Carry out evaluation for tunnel drainage, water-wastewater system and lighting.

(xi) Carry out evaluation for the HV and LV power configuration as well as the provision of signal, control, instruction and monitoring system, station platform screen door system and various auxiliary electromechanical systems.

(xii) Carry out evaluation for the design of major structure, envelope structure, structural waterproofing and structural fire protection of station. Figure 4.27 shows a multi-track station tunnel.

(xiii) Carry out evaluation for the construction excavation method, operation organization and influence of construction on the protection and reinforcement measures as well as safety of adjacent buildings and roads.

(xiv) Carry out design evaluation for the spoil disposal of tunnel, type of tunnel portal, greening protection on side and front

Fig. 4.27 Multi-track Station Tunnel

slopes, water and soil conservation as well as greening and rehabilitation of spoil area.

4.4.9 *Key points of special consultation for subgrade, bridge, culvert and tunnel in cold regions*

(1) Key points of consultation for subgrade

(i) Carry out evaluation for the deformation law under the conditions of seasonal frozen soil and frost heaving as well as thawing settlement, and for the engineering measures satisfying the conditions for ballastless track laying.

(ii) Carry out evaluation for the preloading effect of embankment in the section with mollisol (unsaturated soil) from freezing period to thawing period in winter.

(iii) Carry out evaluation for the thermal insulation and anti-freezing properties and effects of subgrade bed materials.

(iv) Carry out evaluation for the anti-frost-heaving and anti-thawing-settlement properties and effects of materials for replacement of subgrade bed.

(v) Carry out evaluation for the strength and crack resistance performance of subgrade surface asphalt concrete mixed with different admixtures and with different mixture proportions at low temperatures.

(vi) Carry out design evaluation for the replacement with non-frost-heaving soil, laying of thermal insulation material and drainage for the low-filled and shallow-excavated subgrade, as well as for the slope stability of shallow cutting.

(vii) Carry out evaluation for the influence of anti-freezing measures for the improved subgrade soil such as the application of admixture on the porosity, permeability, water content, compactness and unconfined compressive strength of the improved soil.

(viii) Carry out evaluation for the construction machinery and mixing method, paving thickness and times of rolling, optimum water content, admixture content, compaction quality control, winter construction and testing method of improved soil.

(ix) Carry out evaluation for the stable slope ratio and protective measures of cutting slope, as well as the stability, materials, construction technology and testing means of retaining structure considering the design of freezing and thawing cycles.

(2) Key points of consultation for bridge

(i) Carry out evaluation for the adaptability of bridge type, pier type and foundation type adopted in design to differential frost heaving and thawing settlement of subsoil.

(ii) Carry out evaluation for the bridge span, pier type and icebreaking measures adopted in design of the river channels with severe drift ice.

(iii) Carry out evaluation for the physical and mechanical properties of reinforced concrete adopted in the design under low temperature conditions.

(iv) Carry out evaluation for the effect of frost shear force of the pile foundation and open caisson foundation adopted in the design under low temperature conditions.

(v) Carry out evaluation for the construction technology adopted in winter construction of bridge.

(3) Key points of consultation for culvert

(i) Carry out evaluation for the culvert structure type, aperture and engineering measures for prevention of poor drainage caused by ice jamming that are adopted in design.

(ii) Carry out evaluation for the buried depths of culvert foundation adopted in design at different frost heaving levels.

(iii) Carry out evaluation for the measures for reduction of frost heaving force to the culvert base and base sides that are adopted in design.

(iv) Carry out evaluation for the construction technology and measures adopted in winter construction of culvert.

(4) Key points of consultation for tunnel

(i) Carry out evaluation for the position and type of tunnel portal in frozen soil regions adopted in design.

(ii) Carry out evaluation for the form and waterproof performance of lining structure adopted in design, as well as the effects of ice pressure and low temperature on lining.

(iii) Carry out evaluation for the effect of frost heaving soil at the back of the portal wall and wing wall, and for the engineering measures adopted in design.

(iv) Carry out evaluation for the construction technology and measures adopted in winter construction of tunnel.

4.4.10 *Key points of special consultation for land subsidence*

(1) Carry out analysis for the natural and economic environment as well as development in land subsidence zones.

(2) Carry out geological environment evaluation for land subsidence zones.

(3) Carry out analysis for the subsidence mechanism and the natural, artificial and train operation influence factors in land subsidence zones.
(4) Carry out evaluation for the subsidence monitoring standards, technology, requirements and arrangement in land subsidence zones.
(5) Carry out analysis and evaluation for the subsidence monitoring data in land subsidence zones.
(6) Carry out analysis, evaluation and prediction of subsidence stability in land subsidence zones.
(7) Carry out evaluation and countermeasure analysis against subsidence risks in land subsidence zones.

4.4.11 *Key points of special consultation for subsidence in subgrade-bridge zones*

(1) Carry out environmental impact analysis during the project construction period.
(2) Carry out analysis and evaluation for the regional land subsidence and groundwater exploitation, special geological environment and natural consolidation of soil layers.
(3) Carry out analysis for the relationship between urban and township enterprise development.
(4) Carry out analysis and evaluation for the statistics of land subsidence amount.
(5) Carry out analysis and prediction of the land subsidence rate as well as analysis and evaluation of the impact on railway engineering.
(6) Carry out analysis and evaluation for bridge and subgrade subsidence.
(7) Carry out evaluation for the engineering prevention measures and emergency response plan against the hazards of land subsidence.
(8) Carry out analysis and evaluation for the land subsidence observation marks, monitoring period and measures of groundwater exploitation control.

Chapter 5

Cases of Railway Engineering Consultation

5.1 Consultation on Dynamic Responses of Transition Sections Among Subgrade, Bridges, Tunnels and Culverts of a Certain High-speed Railway

5.1.1 *General*

The dynamic performance of a train running on a railway line plays a more and more dominant role in the design of the entire civil works. Due to the effects of such engineering structures as subgrade, bridge, culvert, tunnel and turnout, track rigidity and its deformation are different and uneven along the longitudinal direction of the line. Especially in case of different track foundation structures, the rigidity of the track will change suddenly. For example, dynamic irregularity exists on the connections between the bridge track and the subgrade track at both ends, between the track outside a turnout and the track in turnout section, between the track outside a tunnel and the track in the tunnel and between the ballastless track and the ballasted track, and it is most severe at the track where the bridge abutment joints the subgrade. When a train passes through these sections, due to the abrupt change of track rigidity and deformation, the wheel/rail dynamic interaction may increase, thus accelerating the deformation and damage of the track and affecting the safety and stability of train operation. The faster the running speed is, the greater the wheel/rail dynamic interaction will be.

In order to reduce the wheel/rail dynamic interaction due to the abrupt change of track rigidity and deformation, it is necessary to set transition sections between the tracks with different track foundation. In the past, as the running speed was not high and the axle load was relatively light, the setting of track transition sections did not draw enough attention. With the construction of a large number of high-speed railways in China, and the running speed of trains getting higher and higher, the requirements for track regularity have become stricter and stricter. It is also more important to set the track transition sections. Over the years, because of the limitation of the disciplines, the analysis of train running safety and riding comfort was often carried out for the track, subgrade, bridge or tunnel separately. But in fact, with the improvement of standards for high-speed railways and the increase of running speed of trains, it is inevitable that such transition sections between subgrade and bridge, between subgrade and culvert, between subgrade and tunnel, between subgrade and turnout, between bridge and tunnel, between bridges, between tunnels and between ballasted track and ballastless track are faced with the change of track rigidity, which may result in the irregularity of the line, and further affect the train running safety and riding comfort. Therefore, the analysis of train running safety and riding comfort shall be carried out not only for the tracks, subgrade and bridges, but also the above-mentioned transition sections. That is to say, analysis, research and comprehensive evaluation of dynamic performance simulation must be carried out for the whole line of high-speed railways, so as to ensure the train running safety and riding comfort.

To ensure that the design of the high-speed railway meets the requirements of relevant technical standards, in the engineering consultation of a certain high-speed railway, the dynamic models of the structures of all types of transition sections are established according to the typical design of subgrade, bridges, culverts, tunnels and transition sections of the high-speed railway; the dynamic theory and evaluation standard for track transition sections are established from the perspective of interaction of high-speed train–track transition sections, to analyze and evaluate the dynamic performance of subgrade–bridge, subgrade–tunnel, subgrade–culvert, bridge–tunnel short subgrade and embankment–cutting transition sections, reasonable rigidity of subgrade surface and reasonable matching design of longitudinal rigidity of the transition sections, and control standard of differential settlement.

The analysis contents mainly include the train–track–structure dynamic simulation analysis of subgrade–bridge transition section, subgrade–tunnel transition section, subgrade–turnout transition section, ballasted track–ballastless track transition section, subgrade–culvert transition section, and short subgrade transition sections between bridges, between bridge and tunnel and between tunnels of the high-speed railway. Research on the safety and comfort of a train when it passes through the above-mentioned transition sections is carried out, so as to lay the foundation for comprehensive evaluation of the whole-line dynamic performance of the train operating on the high-speed railway.

5.1.2 *Analysis method and dynamic model*

The rolling stock/track dynamic interaction is the most essential and important problem of the wheel/rail dynamics in the railway transport system. With the development of high-speed railways, the dynamic action of the rolling stock on the track structure has become more severe; meanwhile, the impacts of track structure and its geometric state on the running safety and riding comfort of the train have become more prominent. To analyze and solve these problems, it is not enough to merely take the rolling stock or the track into account separately; instead, the rolling stock and the track must be regarded as an entire interactional coupling system in the research. Currently, the vehicle/track coupling dynamics has become a basic method for the research in China and abroad on the rolling stock/track dynamic interaction. It has been applied to various aspects including the analysis of dynamic impact effect due to minor defects of wheels or rails, research on vibration characteristics and reasonable parameters of the track structure, evaluation of running

Fig. 5.1 Schematic Diagram of Vehicle/Track Spatial Coupling Dynamic Model

safety and riding comfort of the rolling stock, optimization of plane and profile design parameters of the line, and dynamic performance of the train when it passes through a turnout or a transition section.

In the vehicle/track coupling dynamic model shown in Fig. 5.1, the rolling stock is regarded as a multi-rigid-body system with primary suspension and secondary suspension, consisting of car body, frame and wheel set. In the suspension system of the rolling stock, there is linear or nonlinear damping. For the rolling stock, vertical, transverse, bouncing, pitching, side rolling and yawing degrees of freedom of the car body, front and rear frames and wheel sets are taken into account. The normal wheel/rail force is determined according to the Hertz nonlinear elastic contact theory; the tangential creep force is determined according to the Kalker linear creep theory; and the nonlinear correction is carried out according to Shen–Hedrick–Elkin's Theory, so that the creep force at any creep rate and small spin value can be calculated. The rail is regarded as the Bernoulli–Euler Beam on the bearing foundation supported by elastic points. The bearing points of the rail are arranged as per the actual nodal spacing of the fastenings. The degrees of freedom of the rail include the vertical, transverse and rotational degrees of freedom of the left and right rails. The track bed slab is regarded as the bidirectional bending elastic subgrade slab which is supported on the track foundation. The components interact through the linear spring and viscous damping.

Excitation of wheel/rail system can be classified into geometric track irregularity, dynamic track irregularity and minor defects of wheel and rail. In the subgrade–bridge, subgrade–tunnel, subgrade–culvert transition sections and short subgrade transition section between bridges and tunnels, due to the different track foundation structure, it is inevitable that the residual deformation accumulation will occur on the track under the repeated action of the train load and the effects of environmental climate, resulting in settlement difference of the foundation, and causing track irregularity. Supposing that the settlement difference of the track foundation changes evenly in the transition sections, it can be regarded as the geometric track irregularity and be represented by the limit of deformation angle. Besides, in the transition sections, the rigidity of the track will change a lot, and the change of subgrade rigidity is mainly taken into account in the analysis. For different subgrade structures, their characteristics can be represented by corresponding bearing rigidity of subgrade surface. For different structures of transition sections, the rigidity difference of the track foundations is their common characteristic. Five types of dynamic models of the train–track transition sections are shown in Figs. 5.2–5.6.

5.1.3 *Vehicle/track system dynamics*

According to the vehicle/track dynamic theory, the rolling stock subsystem and the track subsystem are organically connected in

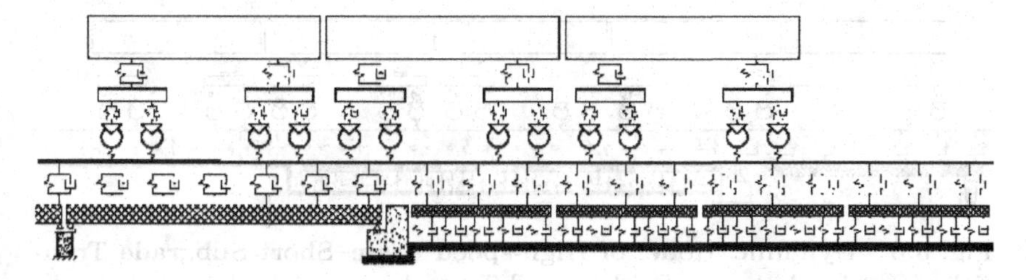

Fig. 5.2 Dynamic Model of High-speed Train–Subgrade–Bridge Transition Section

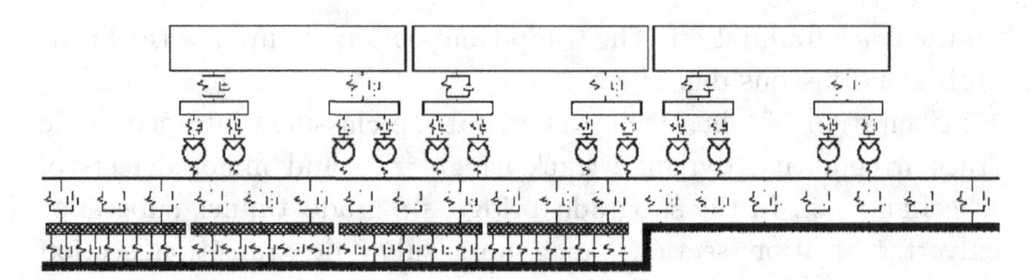

Fig. 5.3 Dynamic Model of High-speed Train–Subgrade–Tunnel Transition Section

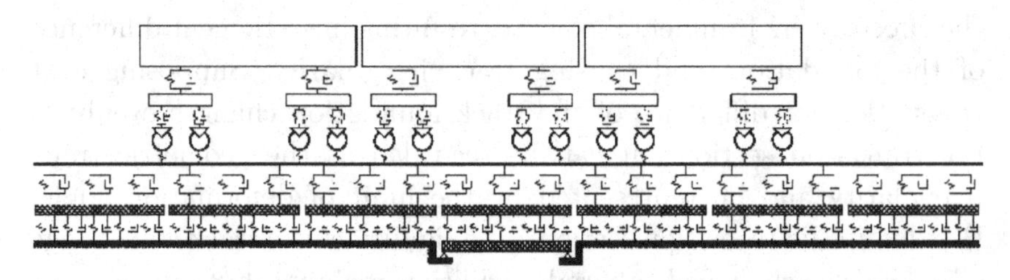

Fig. 5.4 Dynamic Model of High-speed Train–Subgrade–Culvert Transition Section

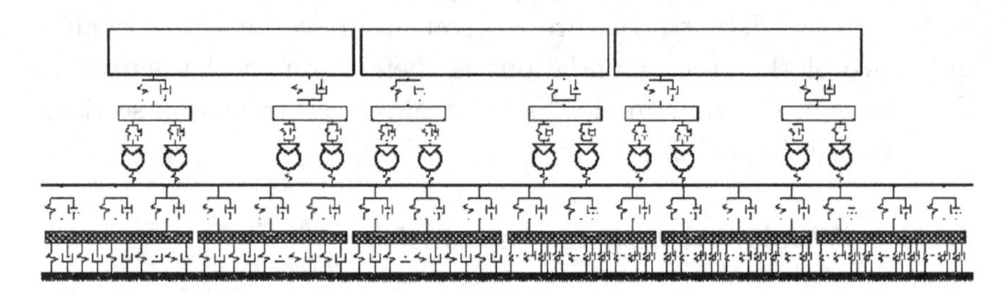

Fig. 5.5 Dynamic Model of High-speed Train–Embankment and Cutting Transition Section

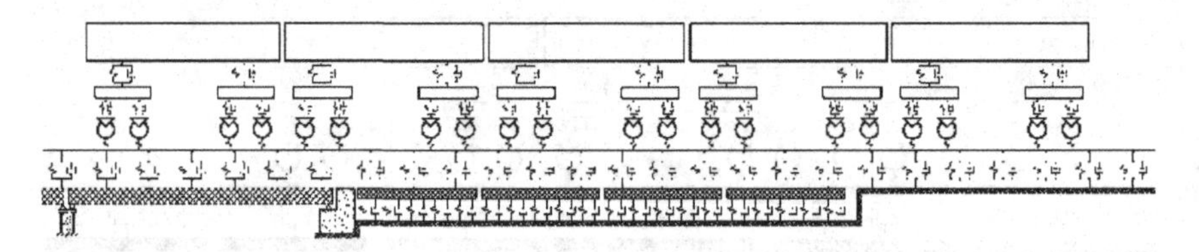

Fig. 5.6 Dynamic Model of High-speed Train–Short Subgrade Transition Section between Bridges and Tunnels

the perspective of system engineering. Based on the rolling stock dynamics, the track structure dynamics and wheel/rail contact geometrical relationship and mechanical relationship, the dynamic analysis model of vehicle/track coupling system is established, taking the geometrical state of the track and the change of parameters of the track structures into full account, so as to carry out the dynamic simulation analysis of the rolling stock and track structures, and evaluate the reasonability of the design parameters of the rolling stock and the track.

(1) Rolling stock dynamic equation
The rolling stock model consists of seven rigid bodies including the car body, frames and wheel sets. Five degrees of freedom including transverse, vertical, side rolling, yawing and pitching degrees of freedom are taken into account for each rigid body. Therefore, each vehicle has 35 degrees of freedom in total, as shown in Table 5.1. The rolling stock dynamic models are shown in Figs. 5.7–5.9. In the figures, the definition of degree of freedom of the frame and wheel set is similar to that of the car body.
(2) Track dynamics
The vibration of the ballastless track is mainly reflected in the vibration of the rail and the track slab (track bed slab). Both

Table 5.1. Degree of Freedom of Dynamic Model of Four-axis Rolling Stock

Degree of Freedom	Longitudinal	Transverse	Vertical	Side Rolling	Yawing	Pitching
Car body	—	Y_c	Z_c	φc	Ψc	βc
Front frame	—	Y_{t1}	Z_{t1}	φ_{t1}	Ψ_{t1}	β_{t1}
Rear frame	—	Y_{t2}	Z_{t2}	φ_{t2}	Ψ_{t2}	β_{t2}
The first wheel set	—	Y_{w1}	Z_{w1}	φ_{w1}	Ψ_{w1}	β_{w1}
The second wheel set	—	Y_{w2}	Z_{w2}	φ_{w2}	Ψ_{w2}	β_{w2}
The third wheel set	—	Y_{w3}	Z_{w3}	φ_{w3}	Ψ_{w3}	β_{w3}
The fourth wheel set	—	Y_{w4}	Z_{w4}	φ_{w4}	Ψ_{w4}	β_{w4}

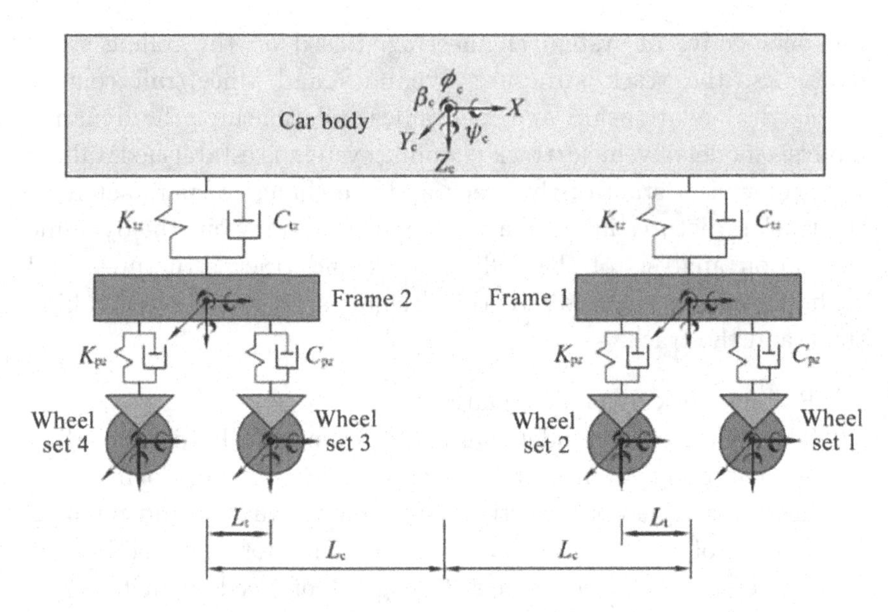

Fig. 5.7 Side View of Dynamic Model of Four-axis Rolling Stock

the left and right rails are regarded as the Euler beam with infinite length on the bearing foundation supported by discrete elastic points, and the vertical, transverse and torsional degrees of freedom are taken into account. The vertical vibration of the track slab (track bed slab) is taken into account as per the rectangular thin plate with equal thickness on the elastic subgrade, and its transverse vibration can be regarded as the rigid body motion. Figures 5.10 and 5.11 are the side view and end view of the dynamic model of the track respectively.

(3) Wheel/rail interaction

Vehicle/track system is a dynamic interacted system, and the wheel/rail relation connects the vehicle subsystem and the track subsystem. According to the dynamic models of rolling stock and track mentioned above, the equation of motion of rolling stock, the equation of motion of track and the wheel/rail interaction force can be derived. Based on the numerical integration method, the dynamic simulation analysis of the rolling stock/track system can be carried out by establishing the computing procedure.

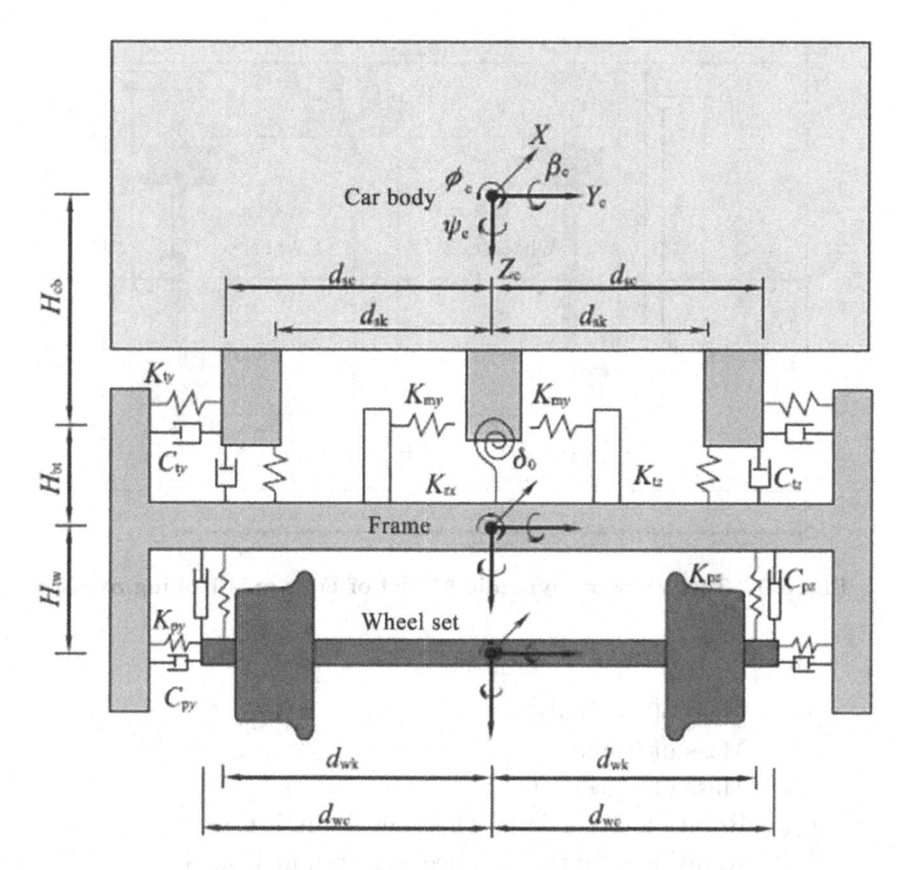

Fig. 5.8 End View of Dynamic Model of Four-axis Rolling Stock

(4) Numerical method for the dynamic equation

In the vehicle/track dynamic model, all degrees of freedom of motion of rolling stock and track are taken into full account; meanwhile, wheel/rail contact geometrical relationship, wheel/rail normal force and wheel/rail tangential creep force are strongly characterized as nonlinear, so the vehicle/track coupling dynamic system is very large and complex. The vehicle/track dynamic equation consists of the equation of motion of rolling stock system and equation of motion of track system.

$$\boldsymbol{M}_{\mathrm{V}}\ddot{\boldsymbol{u}}_{\mathrm{V}} + \boldsymbol{C}_{\mathrm{V}}\acute{\boldsymbol{u}}_{\mathrm{V}} + \boldsymbol{K}_{\mathrm{V}}\boldsymbol{u}_{\mathrm{V}} = \boldsymbol{R}_{\mathrm{V}}$$
$$\boldsymbol{M}_{\mathrm{t}}\ddot{\boldsymbol{u}}_{\mathrm{t}} + \boldsymbol{C}_{\mathrm{t}}\acute{\boldsymbol{u}}_{\mathrm{t}} + \boldsymbol{K}_{\mathrm{t}}\boldsymbol{u}_{\mathrm{t}} = \boldsymbol{R}_{\mathrm{t}}$$

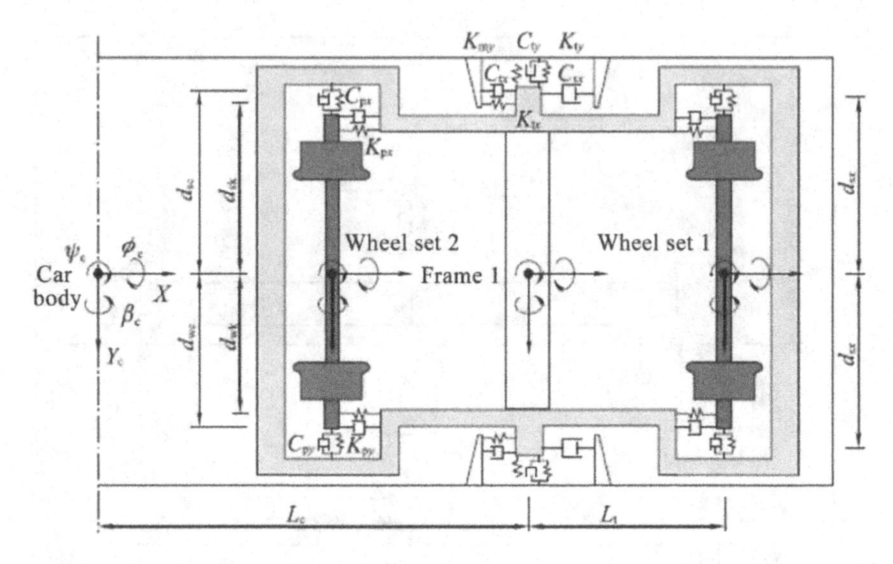

Fig. 5.9 Top View of Dynamic Model of Four-axis Rolling Stock

where

M_c Mass of car body.
M_t Mass of frame.
M_w Mass of wheel set.
I_{wx} Rotational inertia of wheel set around X axis.
I_{wy} Rotational inertia of wheel set around Y axis.
I_{wz} Rotational inertia of wheel set around Z axis.
I_{tx} Rotational inertia of frame around X axis.
I_{ty} Rotational inertia of frame around Y axis.
I_{tz} Rotational inertia of frame around Z axis.
I_{cx} Rotational inertia of car body around X axis.
I_{cy} Rotational inertia of car body around Y axis.
I_{cz} Rotational inertia of car body around Z axis.
K_{tx} Longitudinal rigidity of secondary suspension (at the bogie side).
K_{ty} Transverse rigidity of secondary suspension (at the bogie side).
K_{tz} Vertical rigidity of secondary suspension (at the bogie side).

C_{tx} Longitudinal damping of secondary suspension (at the bogie side).
C_{ty} Transverse damping of secondary suspension (at the bogie side).
C_{tz} Vertical damping of secondary suspension (at the bogie side).
K_{px} Longitudinal rigidity of primary suspension (each axle box).
K_{py} Transverse rigidity of primary suspension (each axle box).
K_{pz} Vertical rigidity of primary suspension (each axle box).
C_{px} Longitudinal damping of primary suspension (each axle box).
C_{py} Transverse damping of primary suspension (each axle box).
C_{pz} Vertical damping of primary suspension (each axle box).
K_{my} Rigidity of lateral stop.
K_{rx} Anti-side-rolling rigidity.
C_{sx} Damping of anti-yaw damper.
H_{cb} Distance between mass center of car body and mass center of bolster.
H_{bt} Distance between mass center of bolster and mass center of frame.
H_{tw} Distance between mass center of frame and mass center of wheel set.
l_{c} Half of length between rolling stock centers.
l_{t} Half of rigid wheelbase of rolling stock.
d_{sk} Half of transverse distance of central spring.
d_{sc} Half of transverse distance of secondary vertical damper.
d_{wk} Half of transverse distance of journal spring.
D_{wc} Half of transverse distance of journal damper.
d_{sx} Half of transverse distance of anti-yaw damper.
δ_0 Secondary lateral stop gap.
X Longitudinal displacement (coordinate).
Y Transverse displacement (coordinate).
Z Vertical displacement (coordinate).

where

$\boldsymbol{M}_{\mathrm{v}}, \boldsymbol{m}_{\mathrm{t}}$	Mass matrixes of rolling stock and track systems.
$\boldsymbol{C}_{\mathrm{v}}, \boldsymbol{C}_{\mathrm{t}}$	Damping matrixes of rolling stock and track systems.
$\boldsymbol{K}_{\mathrm{v}}, \boldsymbol{K}_{\mathrm{t}}$	Rigidity matrixes of rolling stock and track systems.
$\boldsymbol{R}_{\mathrm{v}}, \boldsymbol{R}_{\mathrm{t}}$	Generalized load vectors of rolling stock and track systems.
$\boldsymbol{u}_{\mathrm{v}}, \boldsymbol{u}_{\mathrm{t}}$	Generalized displacement vectors of rolling stock and track systems.
$\boldsymbol{u}_{\mathrm{v}}, \boldsymbol{u}_{\mathrm{t}}$	Generalized speed vectors of rolling stock and track systems.
$\boldsymbol{u}_{\mathrm{v}}, \boldsymbol{u}_{\mathrm{t}}$	Generalized accelerated speed vectors of rolling stock and track systems.

5.1.4 *Rigidity of transition sections and uneven settlement of subgrade*

The problems of track transition sections are mainly reflected in the rigidity difference of the track foundation and the settlement difference of the foundation. To research the longitudinal rigidity

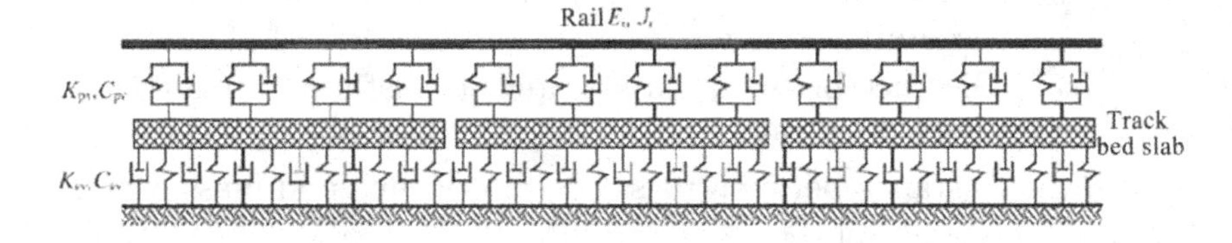

Fig. 5.10 Side View of Dynamic Model of Track

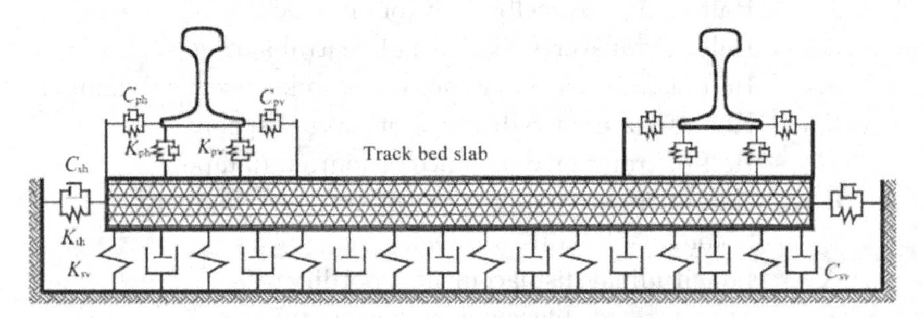

Fig. 5.11 End View of Dynamic Model of Track

matching of subgrade to bridge, tunnel and culvert structures and the dynamic properties of the transition structures, the rigidity range of each part of the line and other relevant parameters should be determined according to the track conditions. According to the design of the high-speed railway, the rigidity of the ballastless track and the ballasted track is analyzed below, and the dynamic analysis scope of uneven settlement is determined with reference to the experience of foreign ballastless track.

(1) Track conditions and basic parameters

The track conditions of the ballastless track of the high-speed railway are designed as per the trans-section continuously welded rail track laid at one time. The rail is designed as 60 kg/m U75V hot-rolled new rail without bolt hole with a length of 100 m. The fastening spacing of the rail is designed as 0.625 m, and the rigidity of the fastenings is 30–50 kN/mm. The track transition section at the starting point and end point of the ballastless track is designed as ballasted track, with the fastening rigidity designed as 30–50 kN/mm, the thickness of the track bed of 0.35 m; the Type III sleeper is adopted, with a length of 2.6 m, and the spacing between sleepers is 0.6 m.

The graded crushed stone base course of subgrade bed of the ballastless track is 0.4 m thick, and it is 0.7 m thick in some individual sections. The thickness of subbase of subgrade bed filled with Group A and Group B fillers is 2.3 m in embankment sections and 1.0–2.3 m in cutting sections. The compaction criteria for base course of subgrade bed, compaction criteria for subbase of subgrade bed and compaction criteria for filling below the subgrade bed are in accordance with *Code for Design of High Speed Railway* (*Trial*). The mechanical property indexes of the bridge–subgrade, culvert–subgrade, tunnel–subgrade, embankment–cutting, short subgrade between two bridges (tunnels) and ballasted track–ballastless track transition sections 3 m below the base course of the subgrade bed are as follows:

Bridge–subgrade transition sections $E = 190$ MPa, $\mu = 0.3$, $E_{\mathrm{vd}} \geq 50$ MPa, $E_{\mathrm{v2}} \geq 120$ MPa

Culvert–subgrade transition sections $E = 190$ MPa, $\mu = 0.3$, $E_{vd} \geq 50$ MPa, $E_{v2} \geq 120$ MPa
General filling sections $E = 120$ MPa, $\mu = 0.3$, $E_{vd} \geq 35$ MPa, $E_{v2} \geq 60$ MPa
Tunnel–subgrade transition sections $E = 190$ MPa, $\mu = 0.3$, $E_{vd} \geq 50$ MPa, $E_{v2} \geq 120$ MPa

(2) Track rigidity
The track transition section is mainly characterized by the rigidity difference resulting from different track foundation structure. The integral rigidity of the track refers to the line rigidity, taking the track and subgrade into comprehensive account.

(3) Analysis of ballasted track rigidity
The rigidity of each part, total rigidity of the line and the vibration mass of the track bed corresponding to one sleeper supporting point can be calculated according to the parameters of the ballasted track of the high-speed railway.

In the transition section between subgrade and bridge or the connection of ballastless track and ballasted track, it is difficult to compact the track bed and the subgrade. Therefore, in the calculation, both the track bed modulus and foundation coefficient are approximated as half of the design values. The calculated rigidity value is basically equivalent to the result obtained from laboratory test under the uncompacted condition of the track bed as well as the measured value at the site.

(4) Analysis of ballastless track rigidity
The elasticity of the slab track mainly comes from the rubber pad under rail and Cement Asphalt (CA) mortar, and the elasticity of bi-block track and long sleeper buried track mainly comes from the rigidity of the rubber pad under rail.

At the connection of ballastless track and ballasted track, under unfavorable conditions, the rigidity of track foundation of the ballastless track is about 3 times of that of the ballasted track, and the integral rigidity of the ballastless track is about 2.5 times of that of the ballasted track.

(5) Bearing rigidity of subgrade surface of ballastless track

Bearing rigidity of subgrade surface refers to the pressure which must be applied on unit area of top surface of subgrade when unit sinking occurs on the top surface of subgrade, and its unit is Pa/m. It represents the elastic characteristics of subgrade and subsoil, and the value is affected by the materials and conditions of subgrade and subsoil.

The structure of bi-block ballastless track on the soil subgrade and the transmission of upper loads are shown in Figs. 5.12 and 5.13, respectively.

To analyze the dynamic effects of bearing rigidity of subgrade surface on the wheel/rail system, the track foundation is simplified as the three-layer structure shown in Fig. 5.14.

Different subgrade structures have different elasticity modulus. In the design of ballastless track of high-speed railway in China, the deformation modulus of graded crushed stone (E_{v2}) is $\geq$120 MPa, and the deformation modulus of Group A and Group B fillers and improved soil (E_{v2}) is $\geq$60 MPa. The elasticity modulus of the intensified subgrade filled with Group A and Group B fillers mixed with 5–8% cement is 800–2,000 MPa; the elasticity modulus of concrete rigid foundation can be 20,000 MPa.

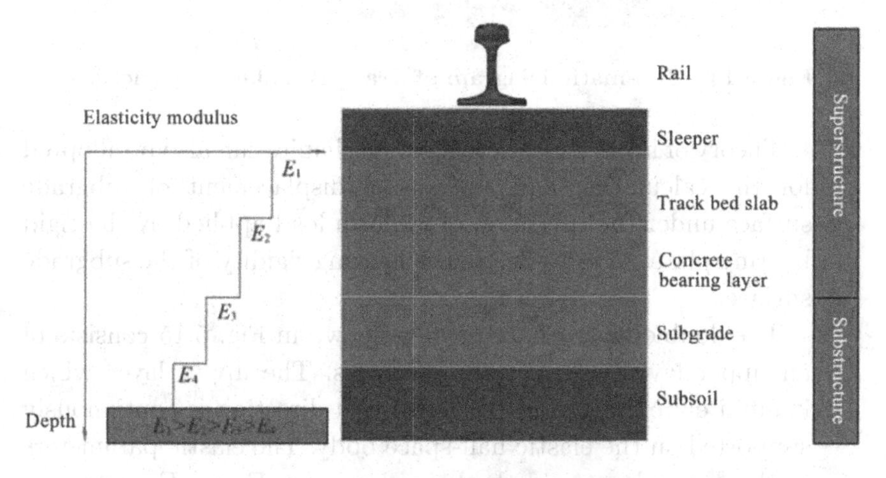

Fig. 5.12 Structural Composition of Ballastless Track

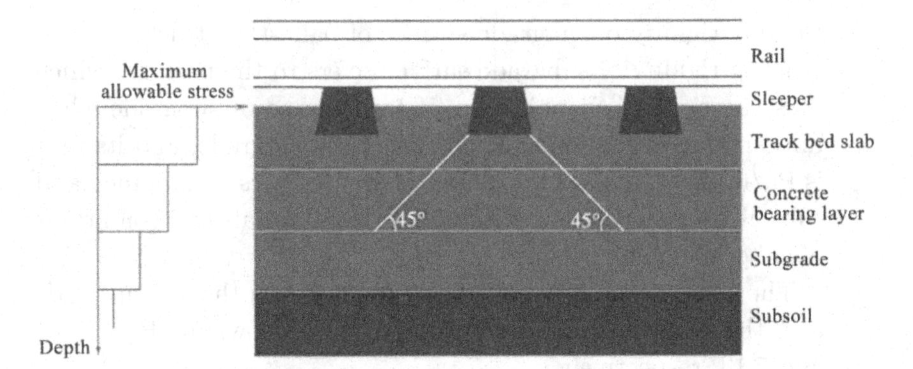

Fig. 5.13 Schematic Diagram for the Downward Diffusion Angle of Upper Loads of Track Structure

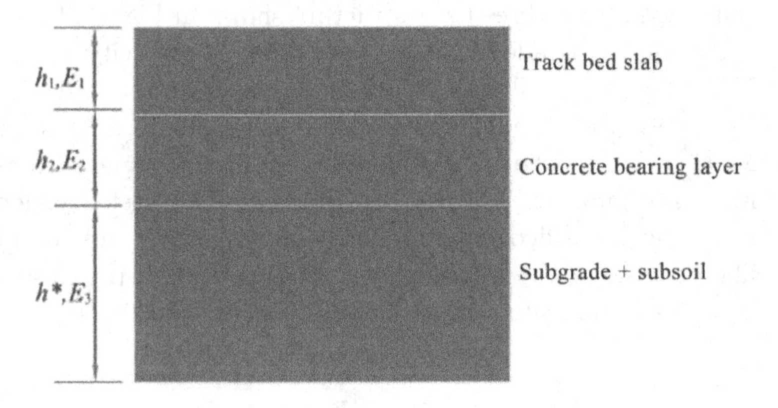

Fig. 5.14 Schematic Diagram of Track Foundation Structure

Theory of layered elastic system mechanics can also be adopted for the calculation and analysis of displacement of subgrade surface under the condition of uniform load applied by the rigid bearing plate, so as to derive the bearing rigidity of the subgrade surface.

The double-layer elastic system shown in Fig. 5.15 consists of an upper layer with certain thickness. The upper layer which is infinitely extended in the horizontal direction is continuously supported on the elastic half-space body. The elastic parameters of the upper layer and the lower layer are E_1, μ_1, E_{21} and μ_{21}, respectively. A circular axial symmetric vertical load effect is

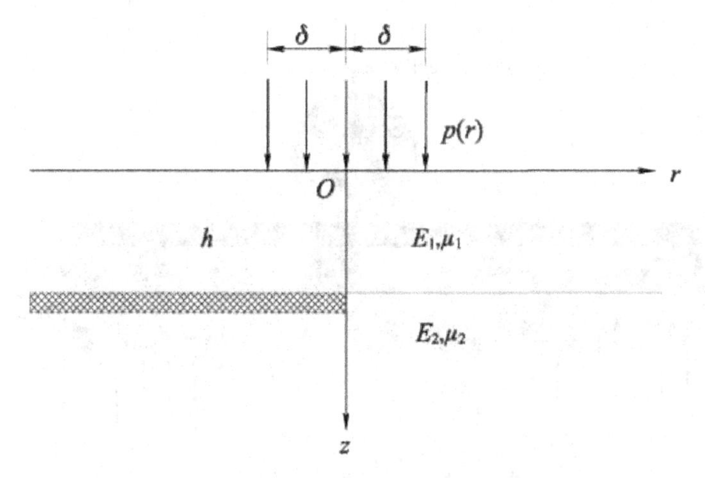

Fig. 5.15 Calculation of Double-layer Elastic System According to Bearing Rigidity of Subgrade Surface

applied on the surface of upper layer. The upper layer represents the base course of the subgrade bed, and the lower layer represents the subbase of the subgrade bed.

(6) Uneven settlement of subgrade of ballastless track

The longitudinal discontinuity of the ballastless track structure in the superstructure of the subsoil foundation along the line will result in peak stress at the discontinuous point of the track structure, as shown in Fig. 5.16. A wide range of subgrade settlement in the shape of long wave will become stable as time goes by. Under the condition of not exceeding the allowable concave deformation, it will not affect the superstructure. However, in order to ensure the comfort level when the train is in operation, the concave deformation of subgrade should meet the requirement for smooth vertical curve of the line. The radius of vertical curve is $r_a \geq 0.4v_E^2$, and the concave deformation of subgrade is $\Delta h = \Delta t^2/4r_a$. See Fig. 5.17.

The subgrade settlement difference in the shape of short wave is unfavorable to the superstructure, and it will affect the running comfort and safety. Therefore, in the design of the transition sections, the continuity of the structures should be maintained as far as possible. Special attention should be paid to the structures

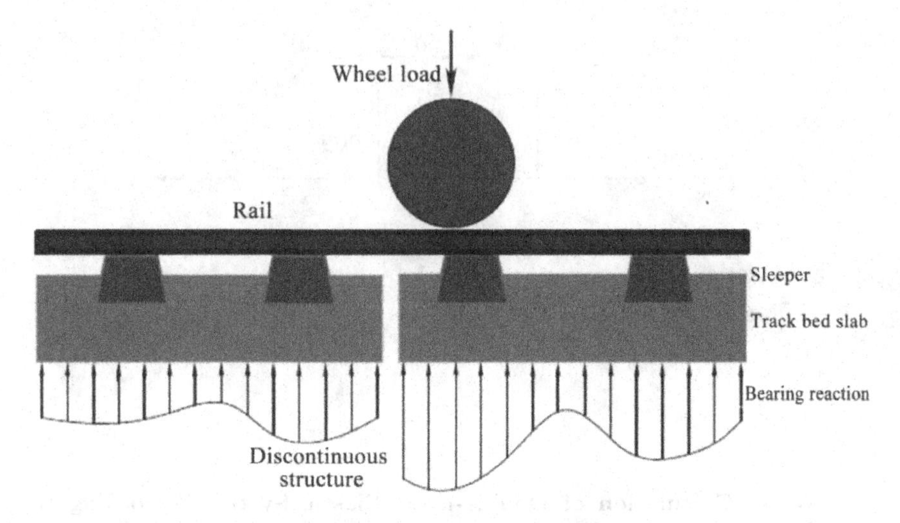

Fig. 5.16 Peak Stress at the Discontinuous Point of the Structure

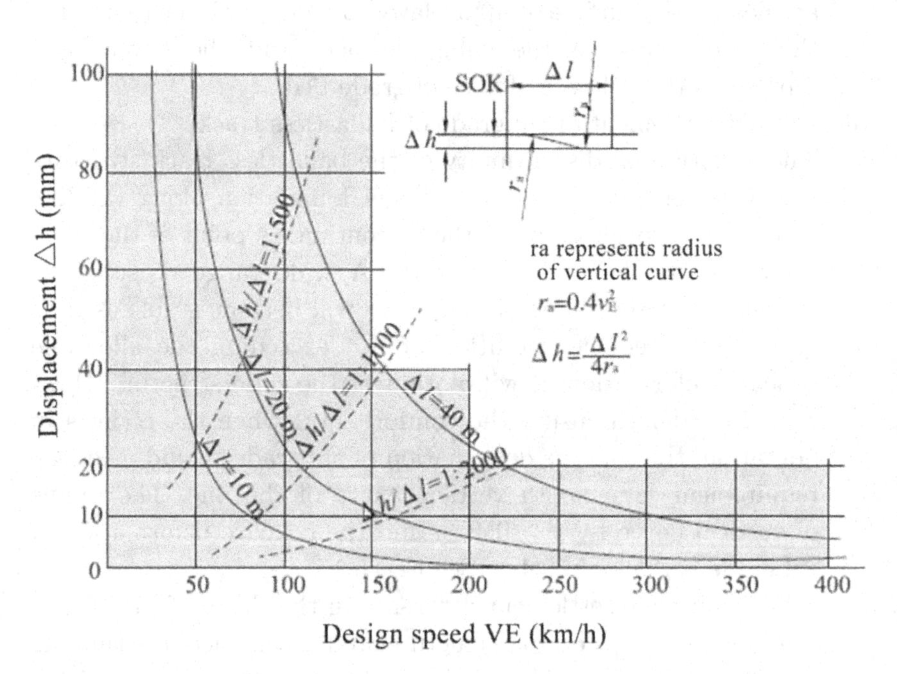

Fig. 5.17 Relationship between Longitudinal Uneven Settlement and Design Speed of the Line

of transition sections between subgrade and bridge, subgrade and tunnel, subgrade and culvert, embankment and cutting or the points of abrupt change of geologic structure. At the beginning of design, relatively large allowable value of settlement and relatively large regulating variable of rail fasteners should not be taken into account. The impacts of settlement difference on the regulating variable of rail fasteners should be minimized, and the limit of foundation settlement difference which has an impact on the rail fasteners should be strictly specified.

In the consultation, dynamic simulation analysis is carried out for the control standards of deformation angle of transition sections of the ballastless track under the operating conditions of the high-speed train.

5.1.5 *Dynamic simulation analysis of typical transition sections*

It is a complex process of wheel/rail dynamic interaction when the train passes through the transition sections. In the process, the parameters of vehicle and track conditions will directly affect the dynamic properties of the wheel/rail system. According to the above-mentioned five transition section models established, three kinds of problems including rigidity difference of track foundation, settlement difference of track foundation and uneven settlement of subgrade can be concluded. The dynamic simulation analysis of the high-speed railway mainly covers the research on the impacts of different vehicle types, different running speed, bearing rigidity of subgrade surface, settlement difference of foundation in transition sections and uneven settlement of subgrade on the wheel/rail interaction, running safety and comfort.

(1) Running conditions

According to the track conditions of the high-speed railway, the following two types of vehicle are selected for the dynamic simulation analysis of typical transition sections:

(i) ICE3 high-speed train, with the running speeds of 200 km, 250 km, 300 km, 350 km and 420 km.

(ii) CRH_3 high-speed train, with the running speeds of 200 km, 250 km, 300 km, 350 km and 420 km.

(2) Evaluation indexes of dynamic properties
Running comfort and safety are evaluated according to Specifications for Strength and Dynamic Performance of Motor Cars of the High-speed Test Trains (95J01-L), Specifications for Strength and Dynamic Performance of Passenger Cars of the High-speed Test Trains (95J01-M), Specifications for Dynamic Performance Evaluation and Accreditation Test of Railway Vehicles (GB5599-85) and Test Verification Method and Evaluation Standard of Dynamic Performance of Railway Locomotives (TB/T2360-93).
Detailed evaluation indexes of wheel/rail dynamic properties include:

(i) Wheel/rail vertical force;
(ii) Rate of wheel load reduction;
(iii) Vertical acceleration of car body and Sperling riding stability index;
(iv) Dynamic stress of subgrade surface.

Vibration acceleration of car body and Sperling riding stability index are two basic indexes used to evaluate the dynamic properties of the rolling stock when it passes through the transition sections. The standards for management of dynamic irregularity and standards for riding stability of rolling stock of the 200–350 km/h track inspection car in China are shown in Tables 5.2 and 5.3.

Table 5.2. Standards for Dynamic Management of Track Irregularity of High-speed Railway

Management Standard	Vertical Vibration Acceleration of Car Body (g)	Transverse Vibration Acceleration of Car Body (g)
Routine maintenance	0.10	0.06
Comfort	0.15	0.10
Emergency repair	0.20	0.15

Table 5.3. Evaluation Rating of Riding Stability of Rolling Stock in China

Stability Grade	Evaluation	Locomotive	Passenger Train	Freight Train
Grade 1	Excellent	<2.75	<2.5	<3.5
Grade 2	Good	2.75–3.10	2.5–2.75	3.5–4.0
Grade 3	Qualified	3.10–3.45	2.75–3.0	4.0–4.25

The dynamic index of the running safety of the rolling stock is mainly the rate of wheel load reduction. According to the Specifications for Dynamic Performance Evaluation and Accreditation Test of Railway Vehicles, the first limit is 0.65 and the second limit is 0.6.

According to the evaluation standards of interaction of wheel/rail system, the wheel/rail interaction force of high-speed passenger train is 170 kN.

(3) Dynamic characteristics of rigidity difference of track foundation and its control standard

If the rigidity difference of the track foundation of the ballastless track is relatively large, the wheel/rail interaction will be quite obvious. The following part introduces the dynamic characteristics resulting from the abrupt change of track foundation rigidity and the arrangement principles of transition section of rigidity. It is assumed that the track foundation rigidity at one side of the line is 15 MN/m, and that at the other side of the line is 60 MN/m. Figures 5.18 and 5.19 show such dynamic responses as wheel/rail vertical force, pressure on sleeper, vertical displacement of rail and vertical vibration acceleration of car body when the high-speed train passes, at a speed of 300 km/h, through the location where the abrupt change of track foundation rigidity occurs.

According to the calculated results, the rigidity difference of track foundation results in the rail deflection difference under the wheel load effect, thus causing the wheel/rail dynamic impact effect and arousing the vibration of the cars. At the location where the abrupt change of track foundation rigidity occurs, the

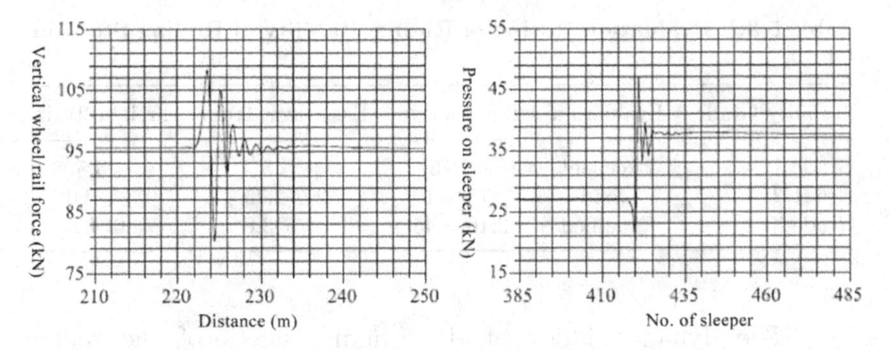

Fig. 5.18 Change of Wheel/Rail Vertical Force and Pressure on Sleeper Resulting from the Rigidity Difference of Track Foundation

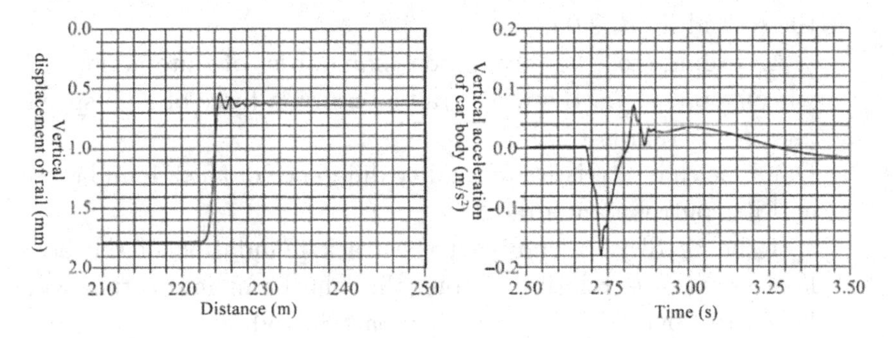

Fig. 5.19 Change of Vertical Displacement of Rail and Vertical Acceleration of Car Body Resulting from Rigidity Difference of Track Foundation

pressure on the sleeper changes sharply, and it is easy to cause such track defects as failure of fastening, suspended or damaged sleeper and track bed settlement under the long-time effects of train load. Once the track defect occurs, the wheel/rail dynamic interaction will be strengthened, resulting in severe impact on driving safety and comfort. Therefore, the transition sections of track rigidity must be arranged.

The rigidity difference of the track foundation can be regarded as dynamic irregularity, and the dynamic method must be adopted for the design and evaluation of transition sections of track rigidity. When a train passes through the track transition section, the requirements for running safety and comfort should be met. Meanwhile, the dynamic effect level of the track structure should be minimized. Dynamic analysis indicates that the "change rate of rail deflection" (slope of dynamic rail deflection curve) is a comprehensive index to evaluate the impacts of wheel/rail dynamic interaction due to the rigidity difference of the track foundation and the impacts of length of the transition sections.

In previous research on the impacts of track rigidity difference, the size of rigidity difference or the concept of "ratio of rigidity" was mostly adopted. In fact, the two indexes cannot accurately reflect the wheel/rail dynamic interaction resulting from the rigidity difference. Detailed description is provided under the following two supposed conditions. Under the first condition, the track foundation rigidity is 5 MN/m and 20 MN/m respectively; under the second condition, the track foundation rigidity is 20 MN/m and 80 MN/m respectively. The ratio of rigidity under both the conditions is 4, while the rigidity difference under the second condition is 60 MN/m, and the difference rigidity under the first condition in 15 MN/m. The wheel/rail dynamic interaction and the rail deflection change rate under the two conditions are given in Table 5.4.

Seen from Table 5.4, the wheel/rail dynamic interaction under the first condition is much larger than under the second condition. The size of rigidity difference reflects the reverse conclusion, and the ratio of rigidity does not reflect the result,

Table 5.4. Comparison of Rail Deflection Change Rate under the Condition of the Same Track Foundation Rigidity Ratio

	Wheel/Rail Interaction Force (kN)	Acceleration of Car Body (m/H)	Rail Deflection Difference (mm)	Change Rate of Rail Deflection (mm/m)
Condition I	112.49	0.408	2.921	2.076
Condition II	107.39	0.145	0.929	1.081

while the "change rate of rail deflection" clearly and accurately reflects this rule.

Seen from Figs. 5.18 and 5.19, under the condition that no transition section is arranged, the effective attenuation distance of the wheel/rail interaction force resulting from the rigidity difference of the track foundation is about 5 m, and the acceleration of the car body reaches 15 m or so. The impacts of length of transition sections of track rigidity on the wheel/rail interaction are evaluated below according to the change rate of rail deflection. Figures 5.20 and 5.21 respectively show the rail deflection curve and curve of change rate of rail deflection under the conditions of different length of transition sections arranged at the location where the abrupt change of track foundation rigidity occurs.

The change rules of change rate of rail deflection, vibration acceleration of car body, wheel/rail dynamic interaction force and pressure on the sleeper under the conditions of different length of transition sections of track foundation rigidity are shown in Fig. 5.22.

According to the results of dynamic simulation calculation, to ensure the high regularity of the track structure of the high-speed

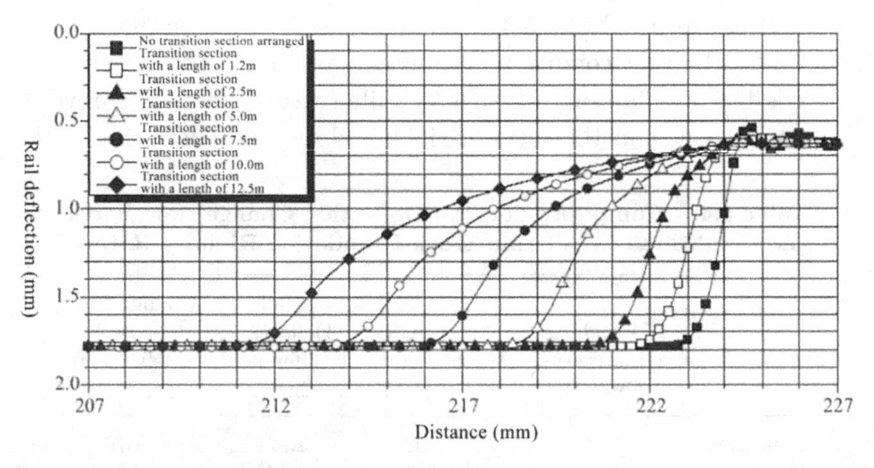

Fig. 5.20 Rail Deflection Curve under the Conditions of Different Length of Transition Sections

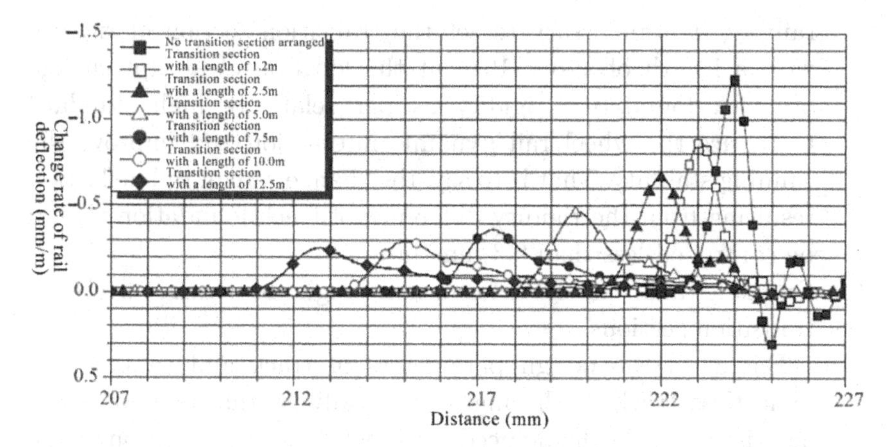

Fig. 5.21 Curve of Change Rate of Rail Deflection under the Conditions of Different Length of Transition Sections

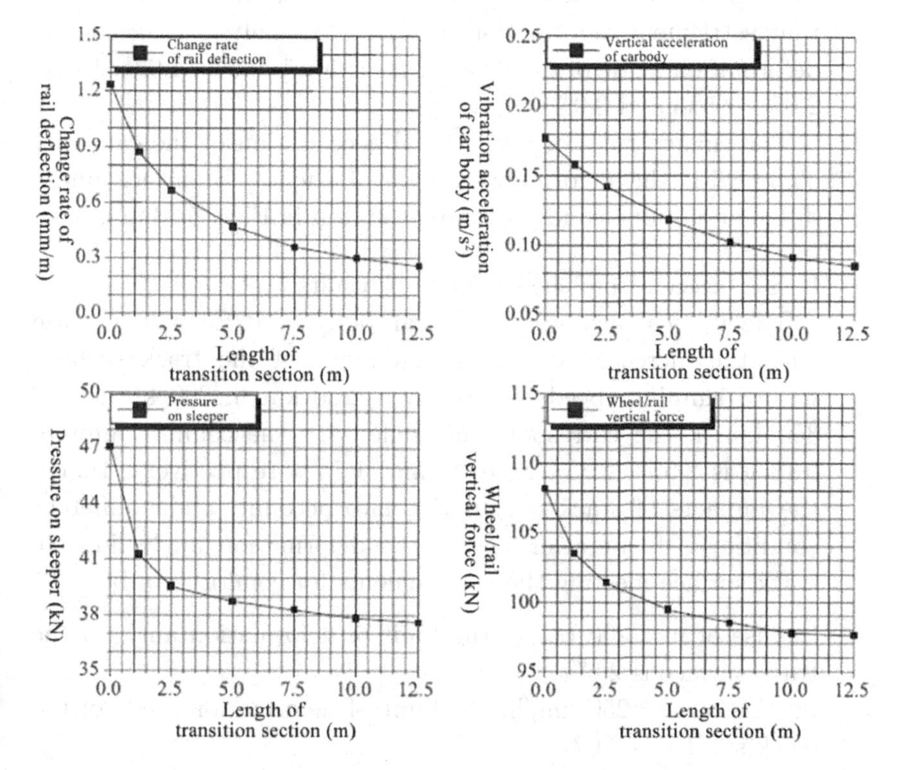

Fig. 5.22 Dynamic Characteristics under Different Length of Transition Section

railway, it is necessary to set the transition section of rigidity with a length of about 10 m at the location of abrupt change of track foundation rigidity as per relatively high standard to reduce the wheel/rail dynamic interaction and improve the running stability. That is to say, the change rate of rail deflection resulting from the rigidity difference of track foundation should be controlled less than 0.3 mm/m.

(4) Dynamic analysis of deformation angle of track surface of transition sections

Based on the design parameters of track and subgrade of ballastless track of the high-speed railway, the dynamic characteristics of the high-speed train when it passes through the transition sections of the ballastless track foundation (especially the ballastless track–ballasted track transition section) shows that settlement deformation occurs. The analysis range of the deformation angle of the transition section is 0.5‰, 1.0‰, 1.5‰, 2.0‰, 2.5‰ and 3.0‰.

The change rules of the vertical acceleration of the car body along with the deformation angle of the track surface under different running conditions are as shown in Figs. 5.23 and 5.24.

It can be seen from the calculation results that:

The vertical acceleration of car body of ICE3 will increase with the increase of deformation angle of the track surface; with a running speed of 200 km/h, 250 km/h, 300 km/h and 350 km/h, the corresponding angle with the comfort limit of 0.13 g is 2.9‰, 2.35‰, 2.05‰ and 1.75‰ respectively. Taking the impacts of track irregularity into account, under different conditions of operating speed, the deformation angle limit of track surface meeting the requirements for comfort can be:

In case of $v = 200$ km/h, the limit of deformation angle of the track surface is 2.5‰.
In case of $v = 250$ km/h, the limit of deformation angle of the track surface is 2.0‰.
In case of $v \geq 300$ km/h, the limit of deformation angle of the track surface is 1.5‰.

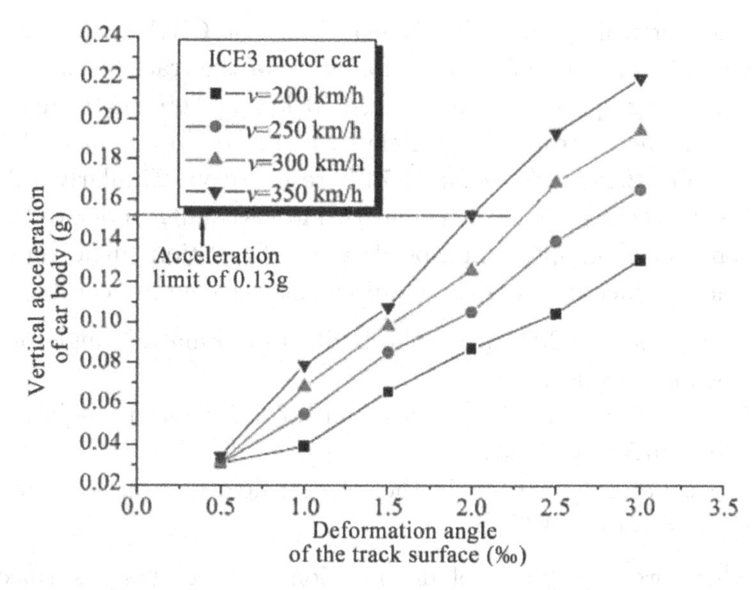

Fig. 5.23 Impacts of Deformation Angle of Track Surface on Acceleration of Car Body

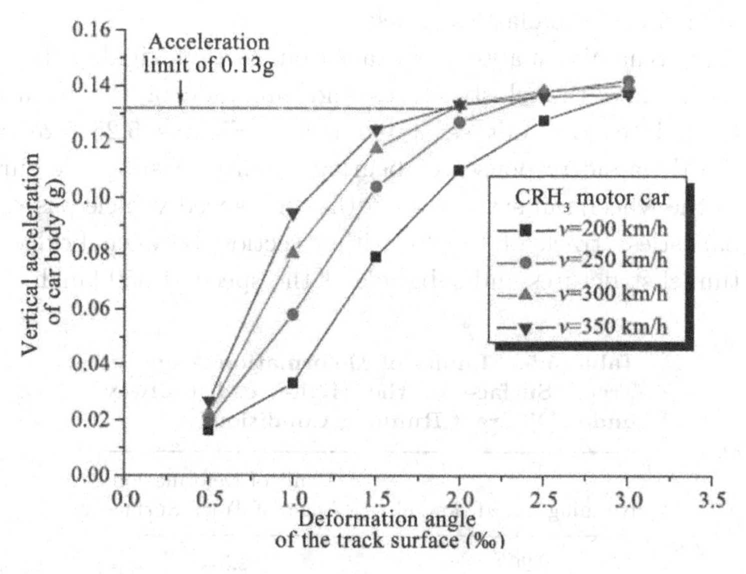

Fig. 5.24 Impacts of Deformation Angle of Track Surface on Acceleration of Car Body

The vertical acceleration of car body of CRH3 will increase with the increase of deformation angle of the track surface; with a running speed of 200 km/h, 250 km/h, 300 km/h and 350 km/h, the corresponding angle with the comfort limit of 0.13 g is 2.6‰, 2.1‰, 1.89‰ and 1.74‰ respectively. Similarly, taking the impacts of track irregularity into account, under different conditions of operating speed, the deformation angle limit of track surface meeting the requirements for comfort can be:

In case of $v = 200$ km/h, the limit of deformation angle of the track surface is 2.5‰.
In case of $v = 250$ km/h, the limit of deformation angle of the track surface is 2.0‰.
In case of $v \geq 300$ km/h, the limit of deformation angle of the track surface is 1.5‰.

Therefore, the limits of deformation angle of track surface of the high-speed railway under different running conditions can be taken as per the values in Table 5.5.

(5) Dynamic analysis of reasonable rigidity of subgrade of transition section of the ballastless track
The dynamic analysis of transition section rigidity between bridge and tunnel structures and subgrade of the ballastless track is carried out as stated below. Figures 5.25–5.28 show the dynamic responses of bearing rigidity of subgrade surface to the wheel/rail system when the high-speed vehicle passes the ballastless track of the transition section between bridge and tunnel structures and subgrade at the speed of 350 km/h.

Table 5.5. Limits of Deformation Angle of Track Surface of the High-speed Railway under Different Running Conditions

Running Speed (km/h)	Limit of Deformation Angle of Track Surface
200	2.5‰
250	2.0‰
≥300	1.5‰

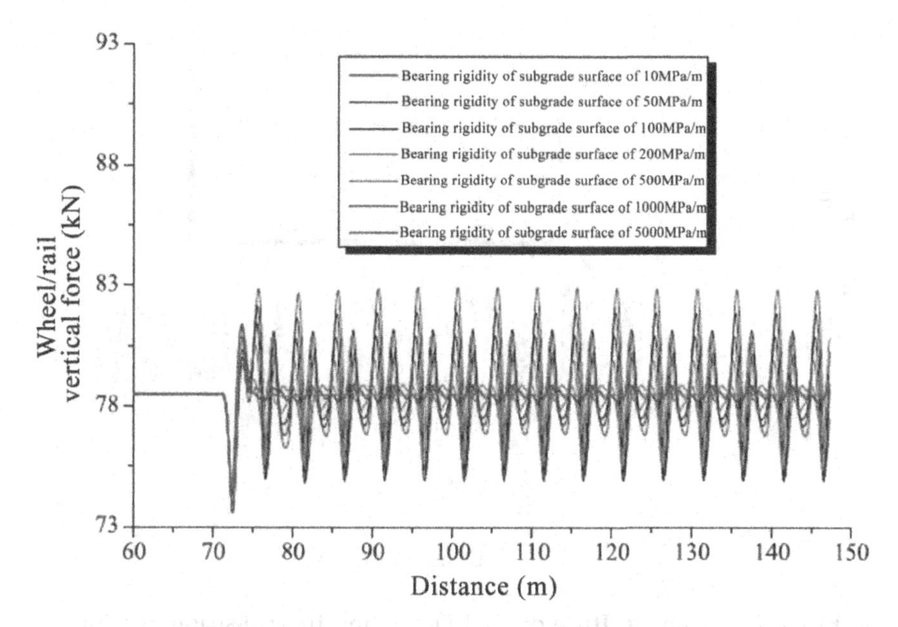

Fig. 5.25 Wheel/Rail Vertical Force in Subgrade–Bridge (Tunnel) Transition Section of Ballastless Track

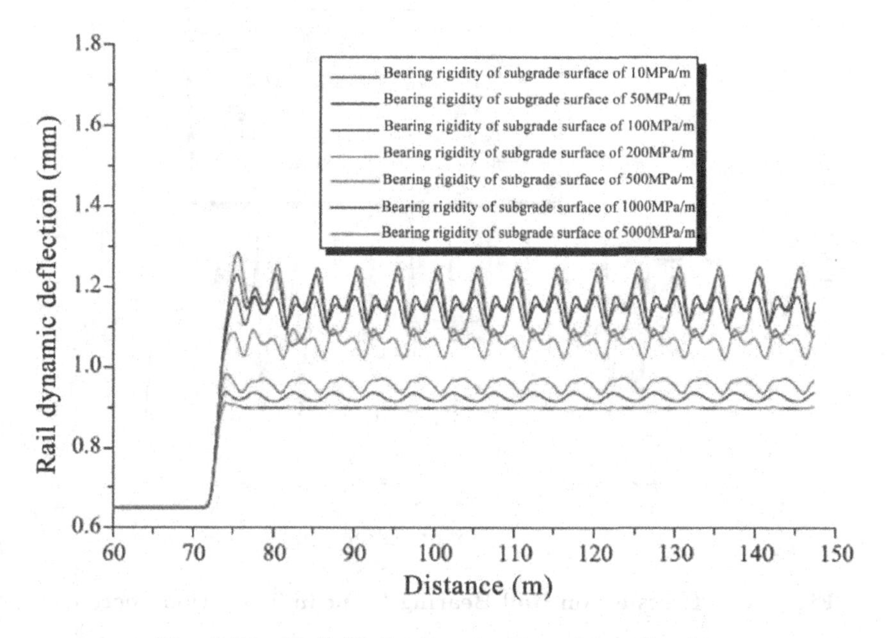

Fig. 5.26 Rail Deflection in Transition Sections

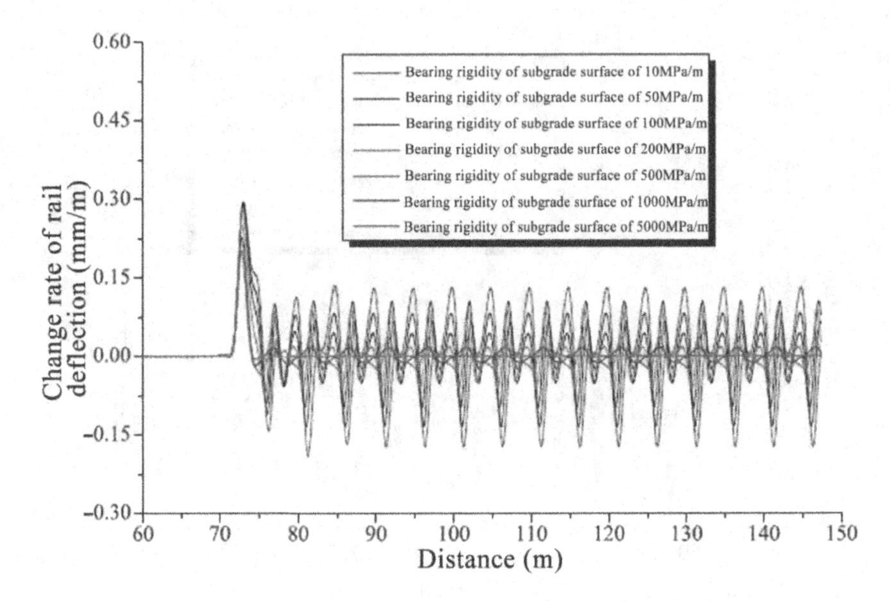

Fig. 5.27 Change Rate of Rail Deflection in Transition Sections

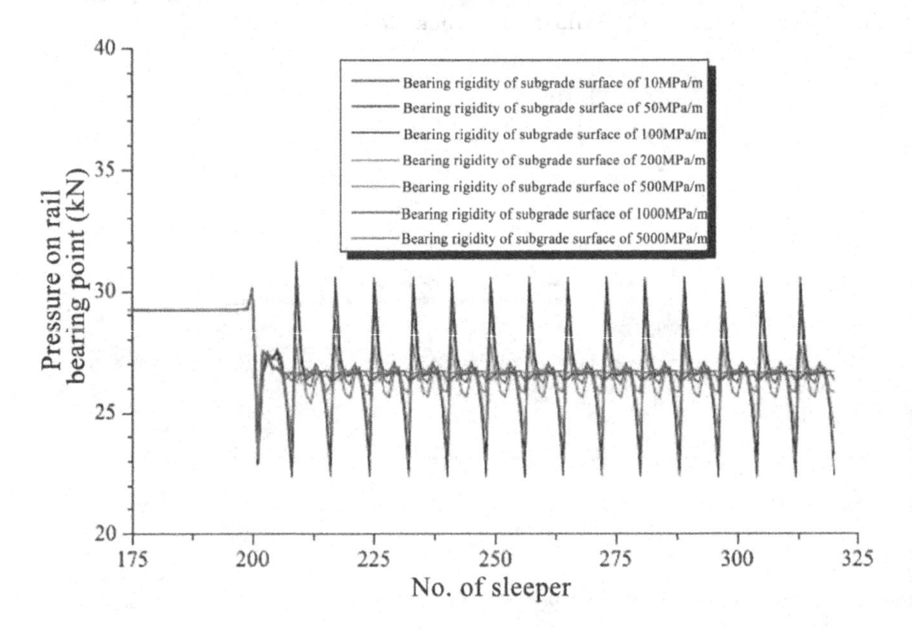

Fig. 5.28 Pressure on Rail Bearing Point in Transition Sections

According to the results of dynamic simulation, for the ballastless track on the subgrade, when the bearing rigidity of subgrade is relatively small, beating of track bed slab will occur under the train load effect. The smaller the bearing rigidity of subgrade is, the more severe beating will be. It is unfavorable for the track structure and the stress and deformation of the subgrade itself, and it will also affect the stability of the track state. When the bearing rigidity of the subgrade surface is more than 1,000 MPa/m, it has little effect on the wheel/rail system. According to the calculation results of wheel/rail interaction force, pressure on the sleeper, rail deflection and change rate of rail deflection, the economically reasonable value of rigidity of subgrade surface in the transition section of ballastless track is 500–1,000 MPa/m. Seen from the test results of bearing rigidity of subgrade transition section of ballastless track for Suining–Chongqing Railway, the bearing rigidity of graded crushed stone subgrade surface is 200–500 MPa/m, and the bearing rigidity of subgrade surface made with 3–10% cement-stabilized graded crushed stones is 800–2,000 MPa/m. Therefore, the subgrade of the transition section of ballastless track should be filled with cement-stabilized graded crushed stones.

(6) Dynamic analysis of uneven settlement limit of subgrade

Supposing that the waveform of uneven settlement of subgrade is a cosine curve, according to the dynamic analysis of uneven settlement of subgrade with combinations of different wave lengths and amplitudes, the uneven settlement limit of subgrade is determined through the research in terms of running comfort. Uneven settlement of subgrade is a kind of partial irregularity of the track, so in the analysis, the impacts of random irregularity are not taken into account. The analytical range of wave length includes 20 m, 30 m and 40 m; the analytical range of settlement amplitude includes 10 mm, 20 mm, 30 mm and 40 mm. Uneven settlement of subgrade is a kind of partial irregularity of the track, so in the analysis, the impacts of random irregularity are not taken into account. The ICE3 vehicle is adopted, with a running speed per hour of 350 km.

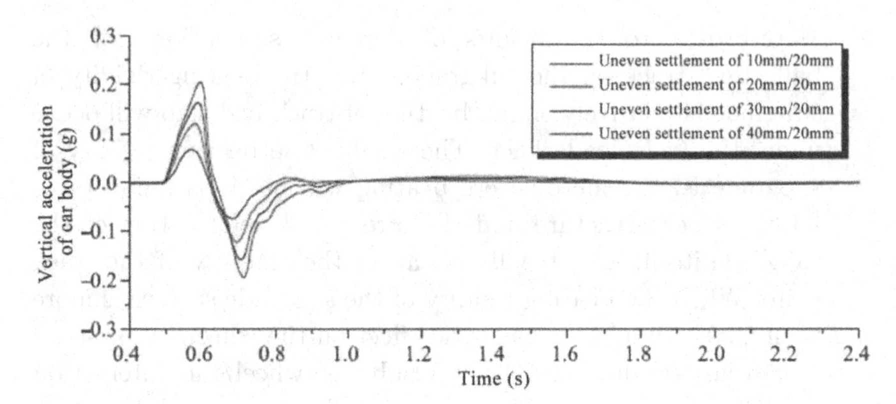

Fig. 5.29 Impacts of Uneven Settlement of Subgrade on Vertical Acceleration of Car Body

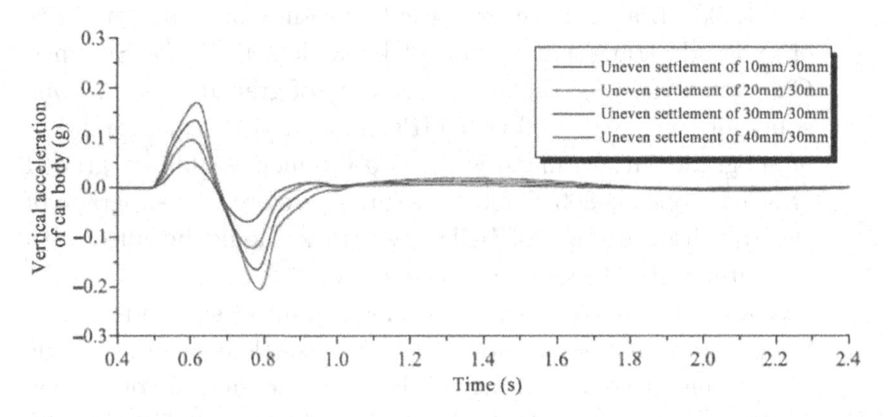

Fig. 5.30 Impacts of Uneven Settlement of Subgrade on Vertical Acceleration of Car Body

Figures 5.29–5.31 show the effect pattern of uneven settlement amplitude of subgrade with wave lengths of 20 m, 30 m and 40 m on the vibration acceleration of car body.

It can be seen from the calculation results of dynamic simulation in Fig. 5.32 that for the comfort standard with the acceleration of car body of 0.13 g, the limit of uneven settlement of subgrade with a wave length of 20 m is 22 mm, that of subgrade with a wave length of 30 m is 29 mm, and that of subgrade with a wave length of 40 m is 36 mm, equivalent to $L/909$, $L/1034$ and

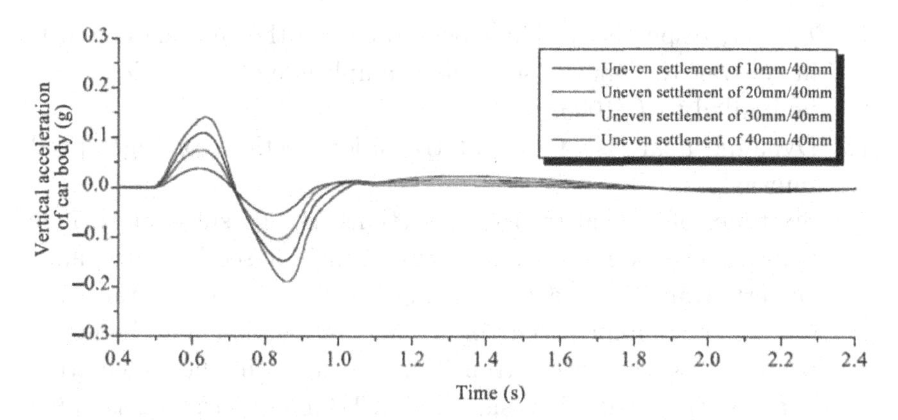

Fig. 5.31 Impacts of Uneven Settlement of Subgrade on Vertical Acceleration of Car Body

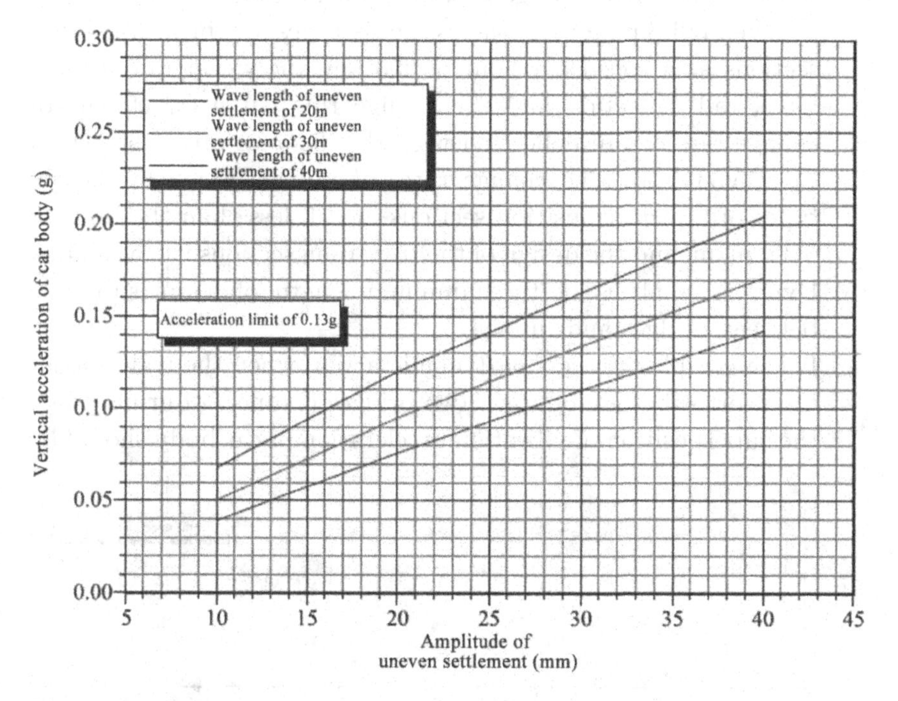

Fig. 5.32 Impacts of Uneven Settlement of Subgrade on Vertical Acceleration of Car Body

$L/1111$, respectively. Therefore, to ensure the running comfort, in general, the uneven settlement amplitude of subgrade should be limited to $L/1000$.

(7) Dynamic analysis of typical transition sections of high-speed railway

Six types of typical transition sections include subgrade–bridge transition section, subgrade–tunnel transition section, subgrade–culvert transition section, bridge–subgrade–bridge transition section, embankment–cutting transition section and ballasted track–ballastless track transition section. In the calculation analysis, the vehicle is designed as ICE3 motor car, with running speeds per hour of 200 km, 250 km, 300 km, 350 km and 420 km. The dynamic response indexes include wheel/rail vertical force, rail deflection and change rate of rail deflection.

When ICE3 motor car passes the six types of typical transition sections at a speed per hour of 350 km, the wheel/rail vertical force, rail deflection and the change rule curve for change of change rate of rail deflection with the distance are as shown in Figs. 5.33–5.50. The change rates of rail deflection of the six types of typical transition sections are all less than the limit of 0.3 mm/m, and the design of the transition sections is reasonable.

(8) Dynamic analysis of the minimum length of short subgrade between bridges and tunnels

It is a complex process of dynamic action when the train passes the short subgrade between two bridges (tunnels), during which the superposition of vibration acceleration of car body should be

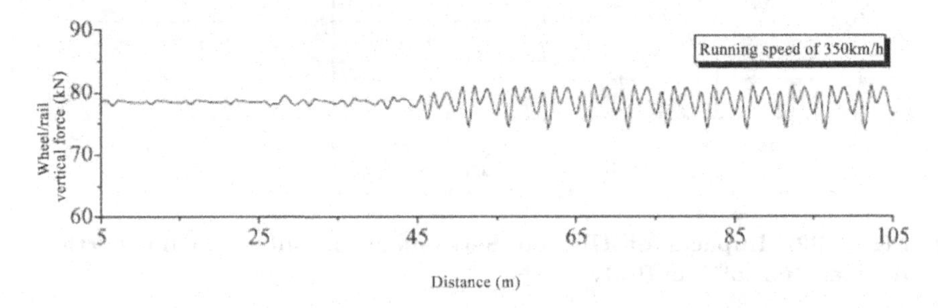

Fig. 5.33 Wheel/Rail Vertical Force in Subgrade–Bridge Transition Section

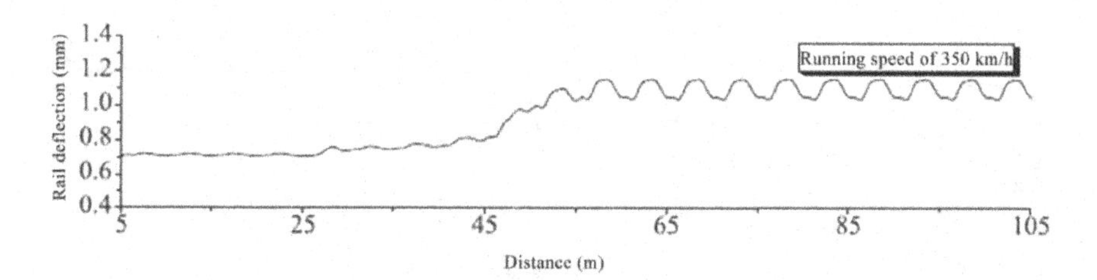

Fig. 5.34 Rail Deflection in Subgrade–Bridge Transition Section

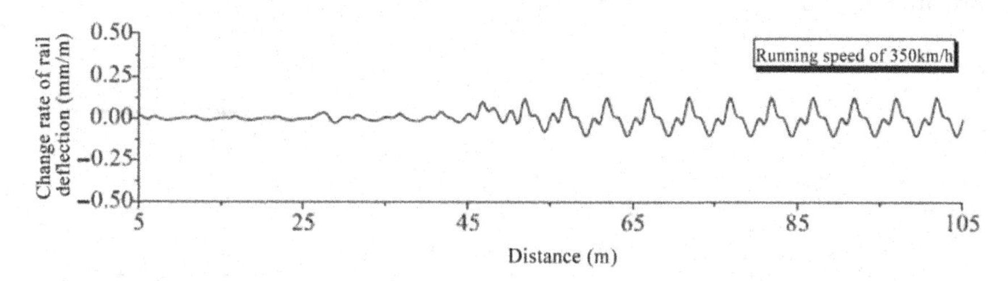

Fig. 5.35 Change Rate of Rail Deflection in Subgrade–Bridge Transition Section

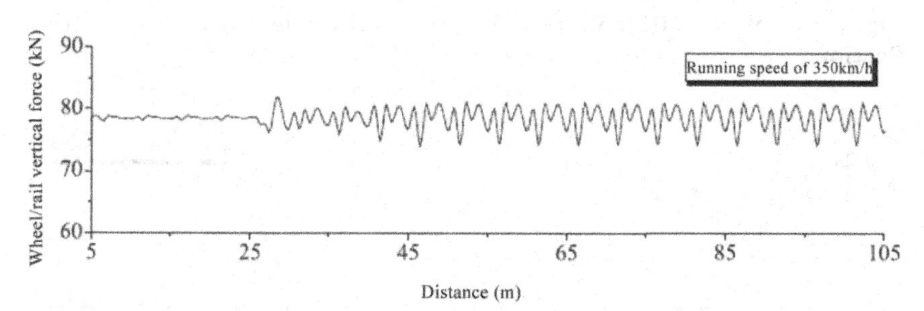

Fig. 5.36 Wheel/Rail Vertical Force in Subgrade–Tunnel Transition Section

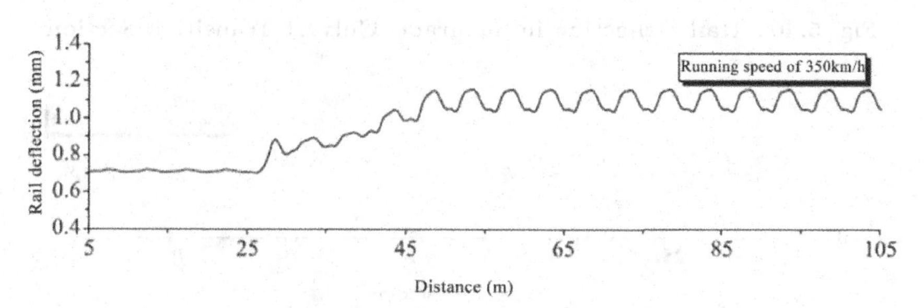

Fig. 5.37 Rail Deflection in Subgrade–Tunnel Transition Section

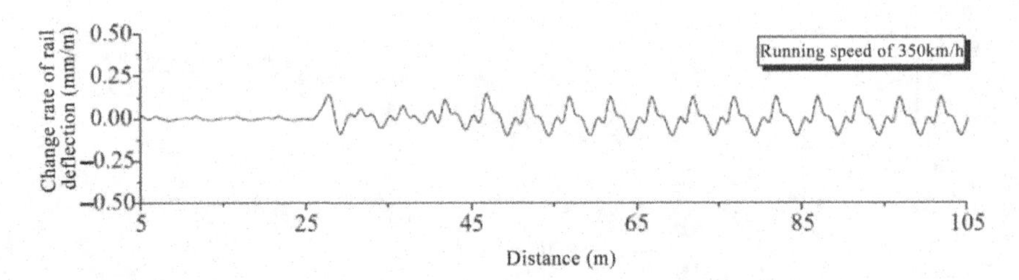

Fig. 5.38 Change Rate of Rail Deflection in Subgrade–Tunnel Transition Section

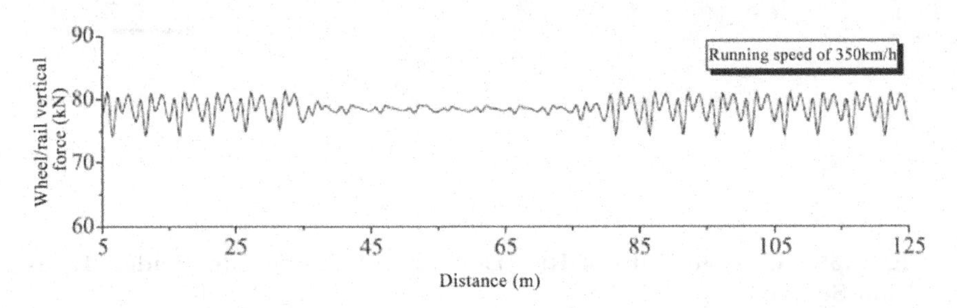

Fig. 5.39 Wheel/Rail Vertical Force in Subgrade–Culvert Transition Section

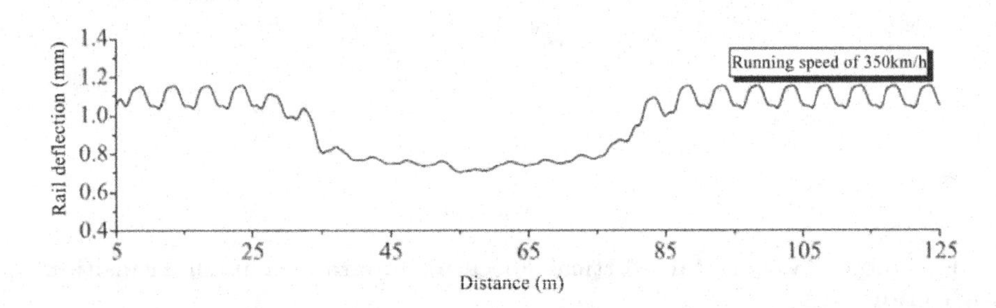

Fig. 5.40 Rail Deflection in Subgrade–Culvert Transition Section

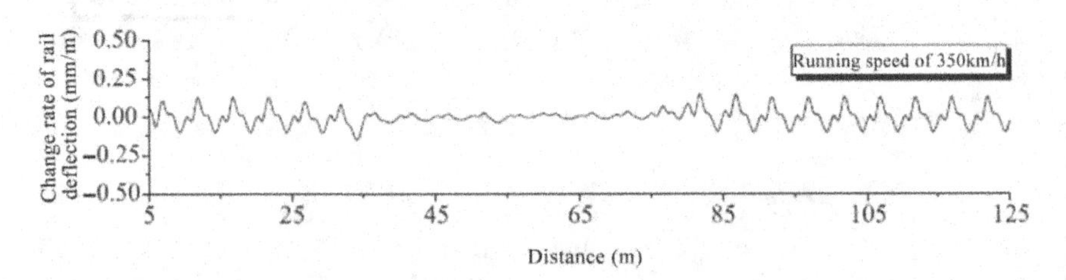

Fig. 5.41 Change Rate of Rail Deflection in Subgrade–Culvert Transition Section

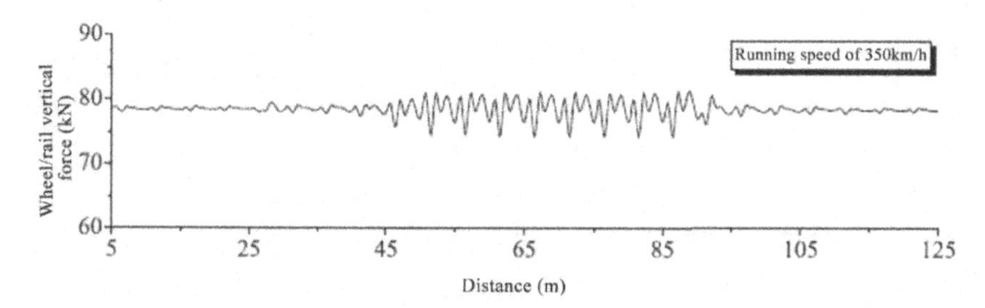

Fig. 5.42 Wheel/Rail Vertical Force in Bridge–Subgrade–Bridge Transition Section

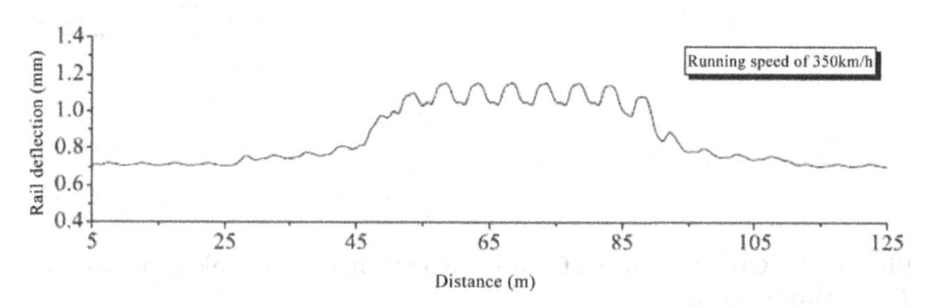

Fig. 5.43 Rail Deflection in Bridge–Subgrade–Bridge Transition Section

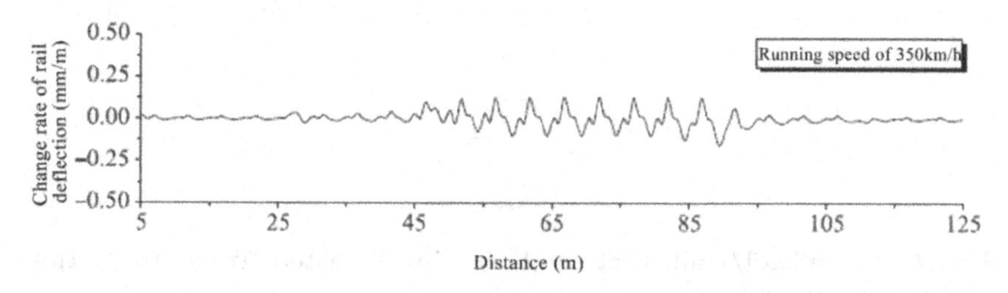

Fig. 5.44 Change Rate of Rail Deflection in Bridge–Subgrade–Bridge Transition Section

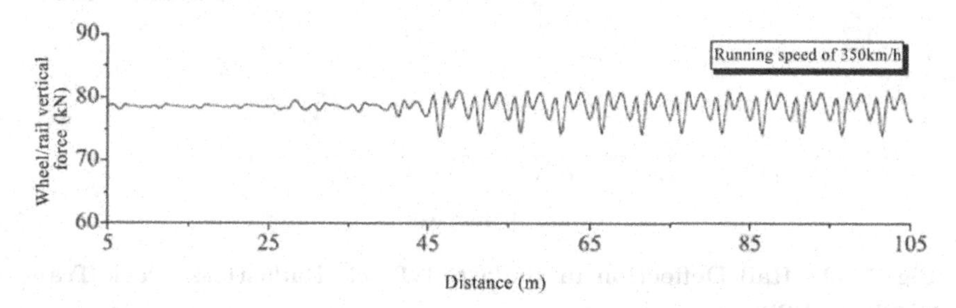

Fig. 5.45 Wheel/Rail Vertical Force in Embankment–Cutting Transition Section

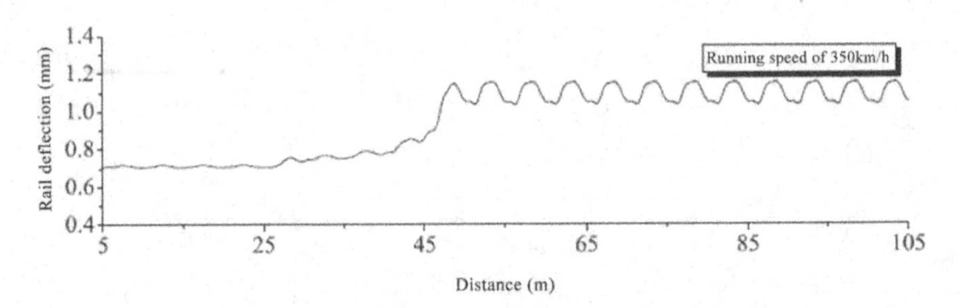

Fig. 5.46 Rail Deflection in Embankment–Cutting Transition Section

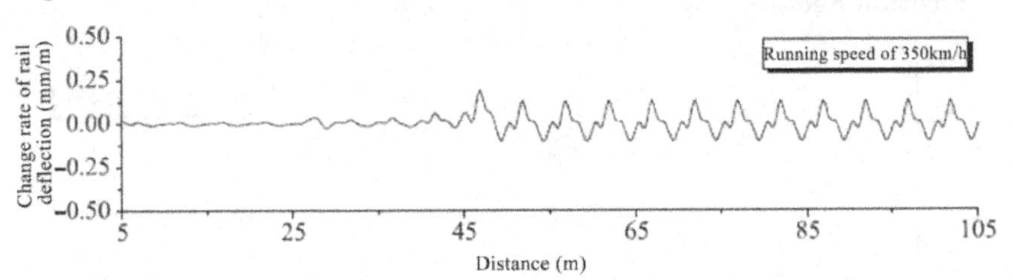

Fig. 5.47 Change Rate of Rail Deflection in Embankment–Cutting Transition Section

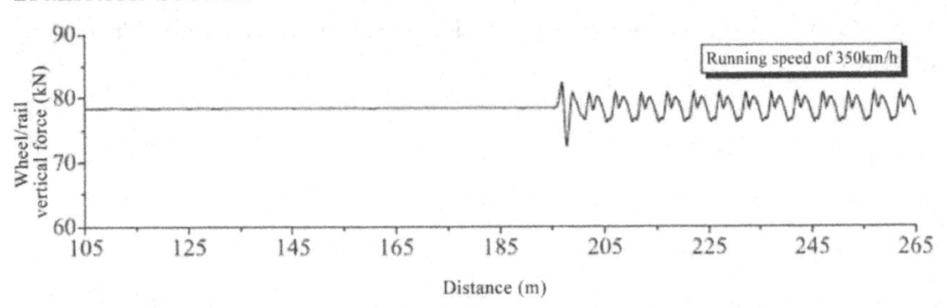

Fig. 5.48 Wheel/Rail Vertical Force in Ballasted Track–Ballastless Track Transition Section

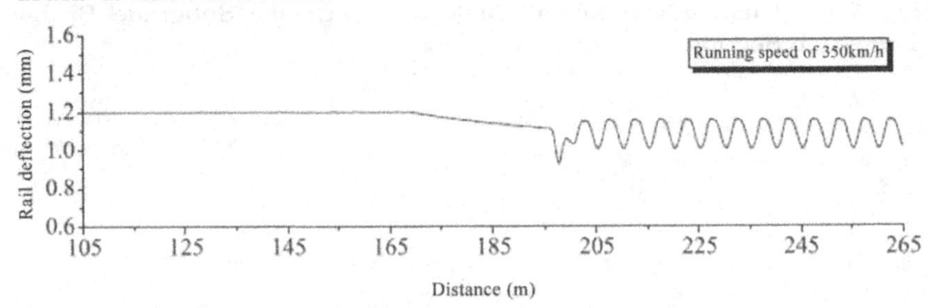

Fig. 5.49 Rail Deflection in Ballasted Track–Ballastless Track Transition Section

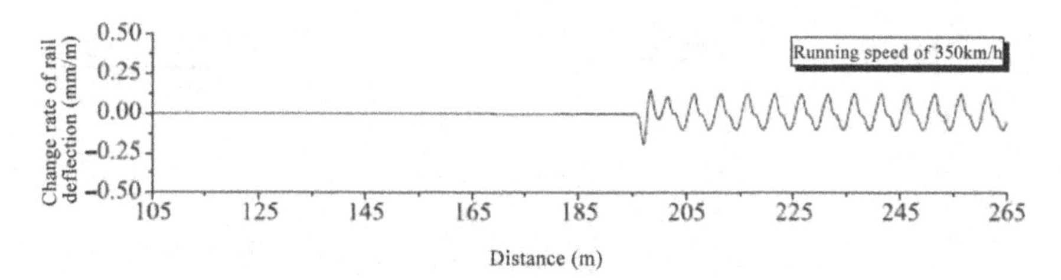

Fig. 5.50 Change Rate of Rail Deflection in Ballasted Track–Ballastless Track Transition Section

mainly taken into account. The value of length of the transition section should be taken as per the limit of deformation angle of track surface. The vibration acceleration of car body of ICE3 motor car and CRH_3 motor car when they pass the short subgrade between the bridges and tunnels at a speed of 200 km, 250 km, 300 km and 350 km are analyzed respectively.

(1) Dynamic analysis when ICE3 motor car passes the short subgrade between bridges and tunnels.
The acceleration responses of car body of ICE3 motor car when it passes the subgrade between bridges and tunnels at different speeds are as shown in Fig. 5.51.
It can be seen from the calculation results that:
   (i) When the ICE3 motor car passes the short subgrade between the bridges and tunnels at a speed of 200 km/h, when the length of subgrade is less than 60 m, the vertical acceleration of car body will increase with the increase of subgrade length; when the length of subgrade is more than 60 m, the vertical acceleration of car body will not change with the increase of subgrade length, so the reasonable length of subgrade is 60 m.
   (ii) When the ICE3 motor car passes the short subgrade between the bridges and tunnels at a speed of 250 km/h, when the length of subgrade is less than 80 m, the vertical acceleration of car body will increase with the increase of subgrade length; when the length of subgrade is more than 80 m, the vertical acceleration of car body

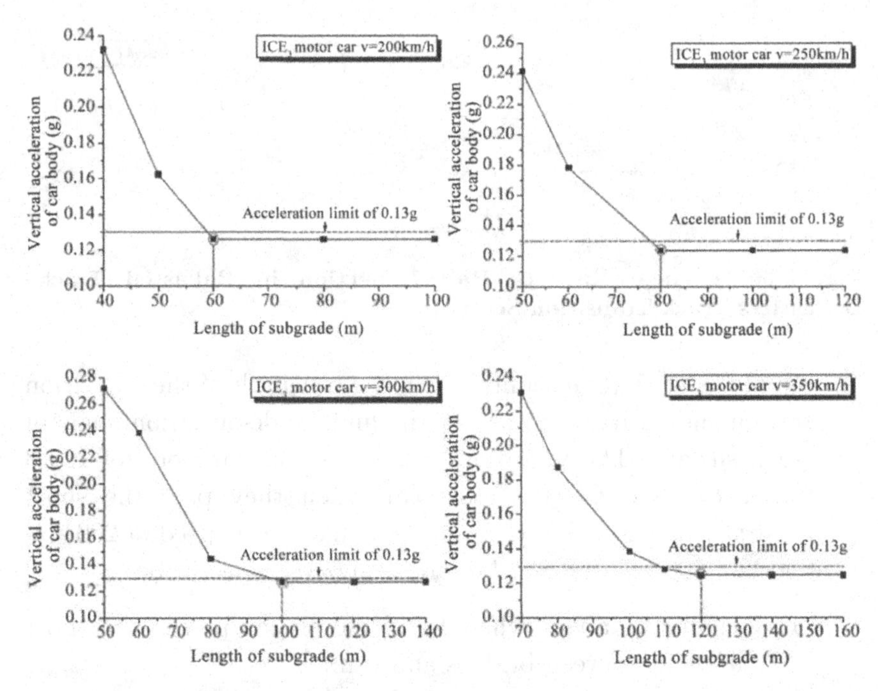

Fig. 5.51 Pattern of Change of Vertical Acceleration of Car Body of ICE3 Motor Car with Length of Subgrade

will not change with the increase of subgrade length, so the reasonable length of subgrade is 80 m.

(iii) When the ICE3 motor car passes the short subgrade between the bridges and tunnels at a speed of 300 km/h, when the length of subgrade is less than 100 m, the vertical acceleration of car body will increase with the increase of subgrade length; when the length of subgrade is more than 100 m, the vertical acceleration of car body will not change with the increase of subgrade length, so the reasonable length of subgrade is 100 m.

(iv) When the ICE3 motor car passes the short subgrade between the bridges and tunnels at a speed of 350 km/h, when the length of subgrade is less than 120 m, the vertical acceleration of car body will increase with the increase of subgrade length; when the length of subgrade is more than 120 m, the vertical acceleration of car body will not change with the increase of subgrade length, so the reasonable length of subgrade is 120 m.

(2) Dynamic analysis when CRH_3 motor car passes the short subgrade between bridges and tunnels

The acceleration responses of car body of CRH_3 motor car when it passes the subgrade between bridges and tunnels with different speed are as shown in Fig. 5.52.

It can be seen from the calculation results that:

(i) When the CRH_3 motor car passes the short subgrade between the bridges and tunnels at a speed of 200 km/h, when the length of subgrade is less than 80 m, the vertical acceleration of car body will increase with the increase of subgrade length; when the length of subgrade is more than 80 m, the vertical acceleration of car body

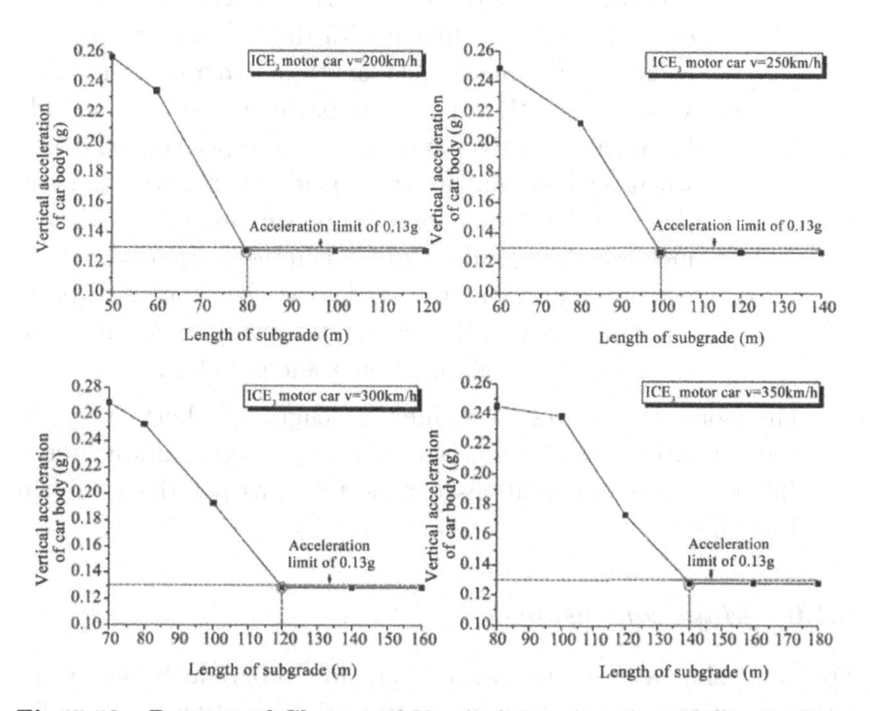

Fig. 5.52 Pattern of Change of Vertical Acceleration of Car Body of CRH_3 Motor Car with Length of Subgrade

will not change with the increase of subgrade length, so the reasonable length of subgrade is 80 m.

(ii) When the CRH_3 motor car passes the short subgrade between the bridges and tunnels at a speed of 250 km/h, when the length of subgrade is less than 100 m, the vertical acceleration of car body will increase with the increase of subgrade length; when the length of subgrade is more than 100 m, the vertical acceleration of car body will not change with the increase of subgrade length, so the reasonable length of subgrade is 100 m.

(iii) When the CRH_3 motor car passes the short subgrade between the bridges and tunnels at a speed of 300 km/h, when the length of subgrade is less than 120 m, the vertical acceleration of car body will increase with the increase of subgrade length; when the length of subgrade is more than 120 m, the vertical acceleration of car body will not change with the increase of subgrade length, so the reasonable length of subgrade is 120 m.

(iv) When the CRH_3 motor car passes the short subgrade between the bridges and tunnels at a speed of 350 km/h, when the length of subgrade is less than 140 m, the vertical acceleration of car body will increase with the increase of subgrade length; when the length of subgrade is more than 120 m, the vertical acceleration of car body will not change with the increase of subgrade length, so the reasonable length of subgrade is 140 m.

Therefore, the values of minimum length of short subgrade between bridges and tunnels of the high-speed railway under different running conditions can be taken as per the values in Table 5.6.

5.1.6 *Main conclusions*

Dynamic simulations are carried out for subgrade–bridge transition section, subgrade–tunnel transition section, subgrade–culvert transition section, bridge–subgrade–bridge transition section,

Table 5.6. Minimum Length of Short Subgrade between Bridges and Tunnels of the High-speed Railway under Different Running Conditions

Running Speed (km/h)	Minimum Length of Short Subgrade between Bridges and Tunnels (m)	Running Speed (km/h)	Minimum Length of Short Subgrade between Bridges and Tunnels (m)
200	80	300	120
250	100	350	140

embankment–cutting transition section and ballasted track–ballastless track transition section of the high-speed railway, and the following conclusions are summed up:

(1) The rigidity difference of the track foundation is regarded as dynamic irregularity, and the dynamic method should be adopted for the design and evaluation of transition sections of track rigidity. Dynamic analysis indicates that the "change rate of rail deflection" (slope of dynamic rail deflection curve) can be adopted for effective evaluation of the impacts of wheel/rail dynamic interaction due to the rigidity difference of the track foundation and the impacts of length of the transition sections. To ensure the high regularity of track structure of the high-speed railway, the change rate of rail deflection resulting from the rigidity difference of track foundation should be controlled to less than 0.3 mm/m.

(2) The foundation settlement difference changes the geometric regularity of the track surface, and it can be characterized by the deformation angle of track surface. The size of the deformation angle of track surface has a significant impact on the running comfort. For the high-speed railway with the design speed per hour of more than 300 km, the limit of deformation angle of track surface should be controlled at 1.5‰.

(3) According to the dynamic analysis of uneven settlement of subgrade with different wave length and different amplitude, to ensure the running comfort, the uneven settlement of subgrade should be limited less than $L/1000$.

(4) According to the dynamic simulation analysis of wheel/rail interaction force, rail deflection and change rate of rail deflection under different conditions of bearing rigidity of subgrade, the economically reasonable value of bearing rigidity of subgrade surface of the ballastless track on the subgrade is 500–1,000 MPa/m. In the construction, 3–10% cement-stabilized graded crushed stones can be adopted for filling, so as to meet the requirement for transition of rigidity of track foundation.

(5) Only when the construction of typical transition sections of the high-speed railway meets the design standards, and under the train load effects, the change rate of rail deflection is less than 0.3 mm/m, the requirements for running safety and comfort can be ensured.

5.2 Consultation on Stress on Girder End of Intercity Railway Simply Supported Girder

5.2.1 *General*

A certain intercity railway is designed with the speed per hour of 200 km, laid with the slab-type ballastless track, and the live load is designed as 0.6UIC.

The whole line is mainly designed as 32 m double-track simply supported girder, and the width of bridge deck structure is 11.6 m. The simply supported girder is designed with the cross section of the circular arc flange plate and inclined web single-box double-chamber box girder, with the girder height at the structure center of 2.2 m, as shown in Fig. 5.53.

5.2.2 *Analytical model of the design structure*

Simulation analysis of the structure is carried out via MIDAS/FEA program. The tetrahedron solid element is adopted, and the girder body is divided into 86,816 elements in total, with consideration of effects of prestressed steel tendon, as shown in Fig. 5.54.

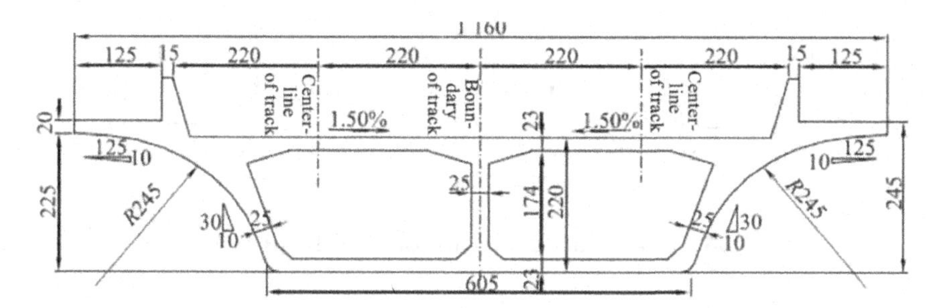

Fig. 5.53 Cross Section of Single-box Double-chamber Box Girder (Unit: cm)

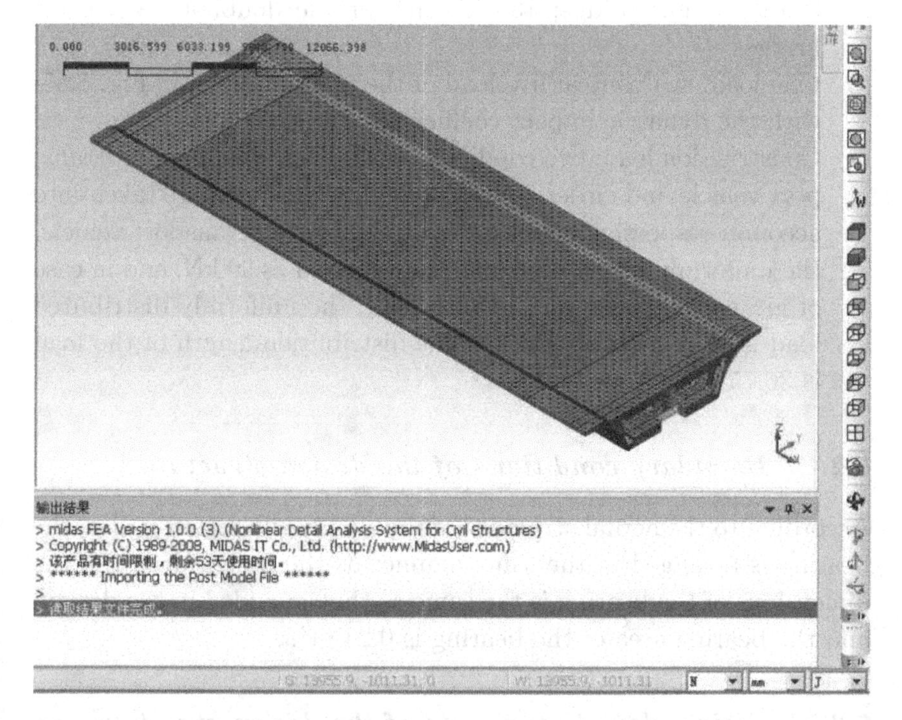

Fig. 5.54 MIDAS/FEA Finite Element Analytical Model of Box Girder

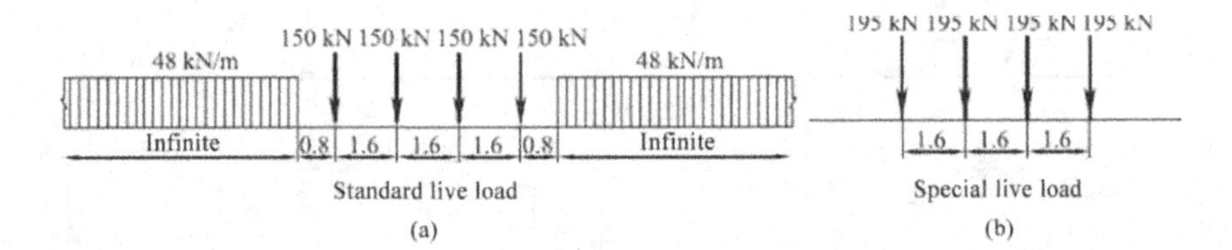

Fig. 5.55 Schematic Diagram of Live Load (Unit: m)

5.2.3 *Calculated load of the design structure*

(1) Dead load: the volume weight of concrete is 26 kN/m^3.
(2) Secondary dead load of bridge deck: 125 kN/m^3 for the double-track straight bridge; 135 kN/m^3 for the double-track curved bridge.
(3) Live load: the vertical live load of the train is shown in Fig. 5.55, with the dynamic impact coefficient of $\varphi = 1.12$.
(4) Construction load: two conditions including no-load girder transport vehicle and girder transport vehicle with load are taken into account, respectively. In case of no-load girder transport vehicle, the uniformly distributed load is calculated as 90 kN, and in case of girder transport vehicle with load, the uniformly distributed load is calculated as 245 kN; the distribution length of the load is 36.75 m.

5.2.4 *Boundary conditions of the design structure*

According to the actual structure of the simply supported girder, the bearing is arranged in the same manner as the design drawing, the dimensions of the bearing is the same as that provided in the design, and the bearing area of the bearing is 0.24 m^2.

5.2.5 *Main calculation results of the design structure*

(1) Operating condition
Self-weight, prestress, secondary dead load and live load are taken into account as the external force of the girder body. According to the influence line for reaction, the location of live

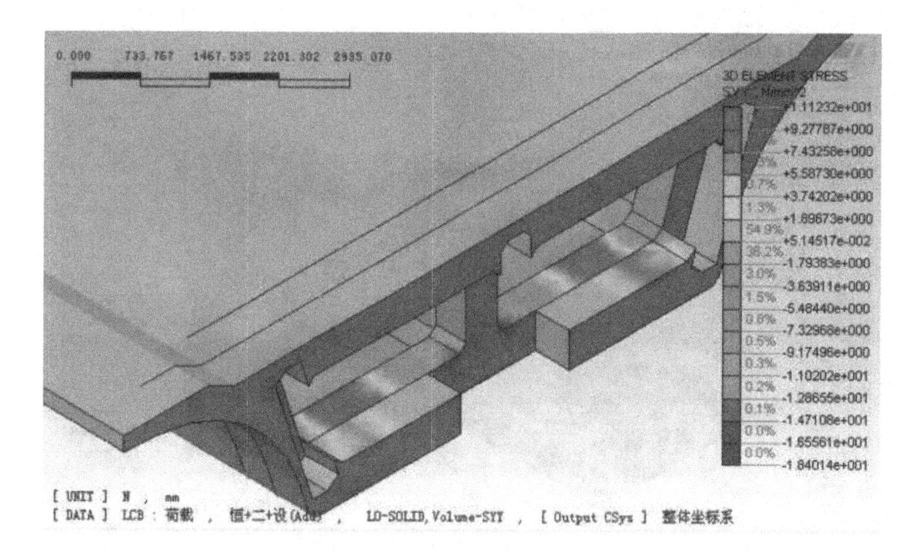

Fig. 5.56 Stress Contour at Girder End under Operating Condition

load action is the location where the maximum bearing reaction exists.

In operating condition, the transverse tensile stress level at the bearing location at the end of the bottom slab is 11.1 MPa. See Fig. 5.56 for the stress contour.

(2) Structure stress under girder transport condition

Self-weight, prestress and design load of girder transport vehicle are taken into account as the external force of the girder body. According to the influence line for reaction, the location of load action of girder transport vehicle is the location where the maximum bearing reaction exists.

In girder transport condition, the relatively large transverse tensile stress exists at the bearing location at the end of the bottom slab, with the stress level of 10.4 MPa. See Fig. 5.57 for the stress contour.

5.2.6 *Calculation conclusions of the design structure*

According to the analysis results of local stress of the girder body, under the load effect of girder transport vehicle, the transverse

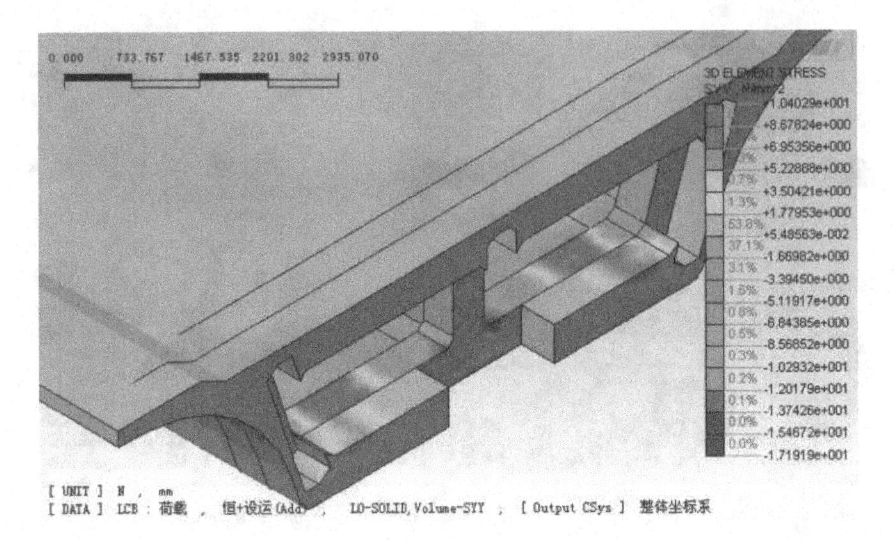

Fig. 5.57 Stress Contour at Girder End under Girder Transport Condition

tensile stress at the bearing location of the bottom slab reaches 10.4 MPa; under operating load effect, the transverse tensile stress at the bearing location of the bottom slab reaches 11.1 MPa. The tensile stress of the concrete exceeds the design limit, which will result in local cracks on the bottom slab.

5.2.7 *Reinforcement scheme*

As the local concrete stress at the girder end of the design structure exceeds the design limit, it is proposed to set the diaphragm at the girder end to improve the local stress of the girder end. Considering that some designed box girders have been prefabricated, to add the diaphragm to the end of the prefabricated box girder, it is difficult to carry out concrete pouring and vibration for the diaphragm top. Besides, due to the gravity and concrete shrinkage, it is difficult to ensure that the diaphragm top is closely attached to the top slab of the box girder as well as the force transmission between them. Therefore, the scheme of arranging the half diaphragm which is easy for construction and quality control needs to be taken into account for comparison.

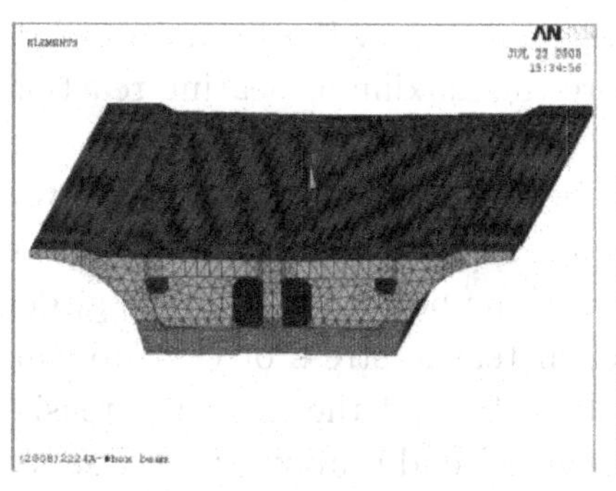

(a) Full diaphragm scheme

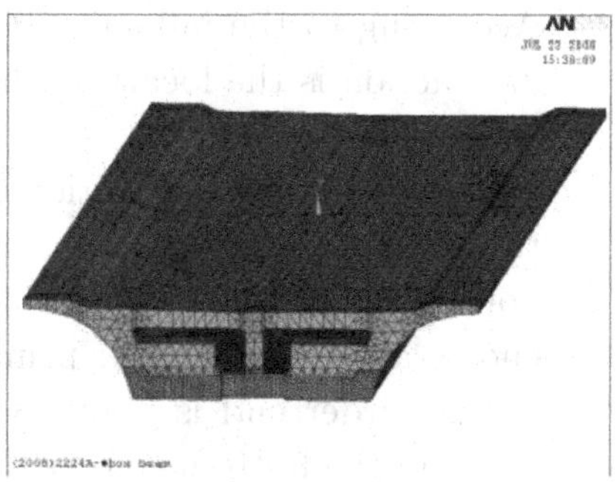

(b) Half diaphragm scheme

Fig. 5.58 ANSYS Finite Element Analytical Model of Box Girder

5.2.8 *Analytical model of the structure in the reinforcement scheme*

The calculation model is established through ANSYS, a large-scale finite element software. SOLID45 solid element is adopted as the girder body element, and LINK8 bar element is adopted as the prestressed steel strand. The model is divided into 202,688 SOLID45 solid elements and 824 LINK8 bar elements in total. See Fig. 5.58(a) for the element model of full diaphragm scheme, and see Fig. 5.58(b) for the element model of half diaphragm scheme.

5.2.9 *Boundary condition of the reinforcement scheme*

The simulation of boundary condition is carried out according to the method of four-fulcrum constraint of the actual structure of the simply supported girder. To reduce the stress concentration due to the fulcrum constraint, the constraint is applied on the bearing bottom of steel plate, in the manner of single-point constraint.

5.2.10 *Main calculation results of full diaphragm reinforcement scheme*

(1) Operating condition

Self-weight, prestress, secondary dead load and live load are taken into account as the external force of the girder body.

According to the influence line for reaction, the location of live load action is the location where the maximum bearing reaction exists.

Under operating conditions, the transverse normal stress of the top slab at the girder end is about 1.51 MPa, and the transverse normal stress of the top surface of the bottom slab at the girder end is about 4.25 MPa; principal tensile stress of the top slab at the girder end is about 3.20 MPa, and the principal tensile stress of the bottom slab at the girder end is about 4.8 MPa. See Fig. 5.59 for the stress contour.

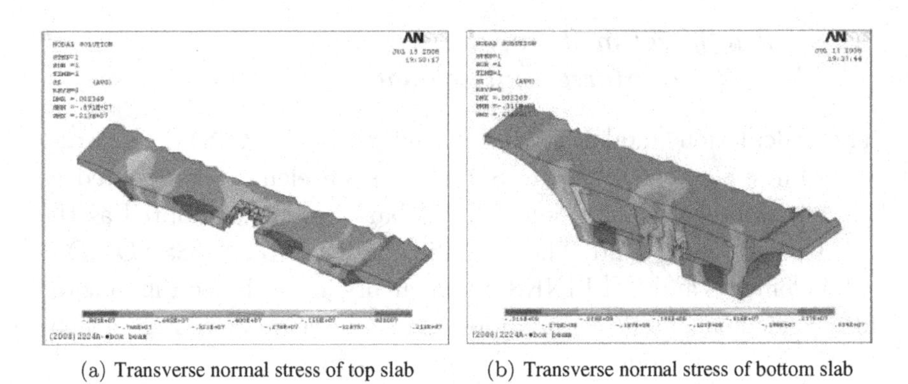

(a) Transverse normal stress of top slab (b) Transverse normal stress of bottom slab

(c) Principal tensile stress of top slab (d) Principal tensile stress of bottom slab

Fig. 5.59 Stress Contour at Girder End under Operating Condition

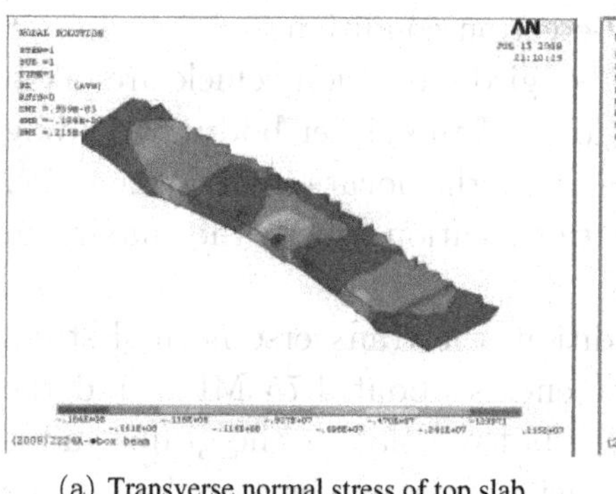

(a) Transverse normal stress of top slab

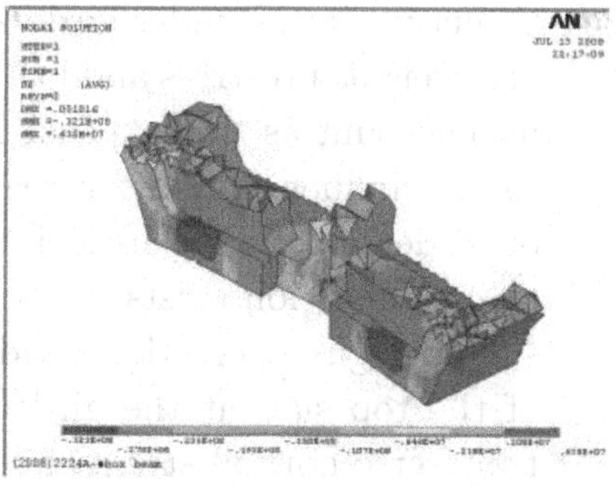

(b) Transverse normal stress of bottom slab

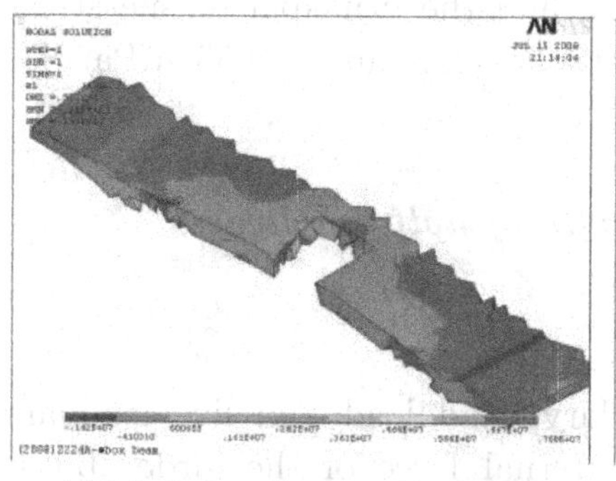

(c) Principal tensile stress of top surface

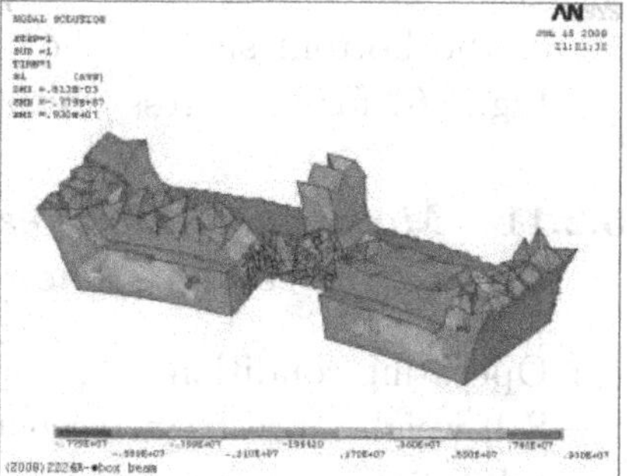

(d) Principal tensile stress of bottom slab

Fig. 5.60 Stress Contour at Girder End under Girder Transport Condition

(2) Structure stress under girder transport condition
Self-weight, prestress and load of girder transport vehicle are taken into account as the external force of the girder body. According to the influence line for reaction, the location of load action of girder transport vehicle is the location where the maximum bearing reaction exists.

Under girder transport condition, the transverse normal stress of the top slab at the girder end is about 1.75 MPa, and the transverse normal stress of the bottom slab at the girder end is about 4.20 MPa; principal tensile stress of the top slab at the girder end is about 2.62 MPa, and the principal tensile stress of the bottom slab at the girder end is about 5.50 MPa. See Fig. 5.60 for the stress contour.

(3) Structure stress under girder erection condition
Self-weight, prestress and load of girder erection vehicle are taken into account as the external force of the girder body. According to the influence line for reaction, the location of load action of girder erection vehicle is the location where the maximum bearing reaction exists.

Under girder erection condition, the transverse normal stress of the top slab at the girder end is about 4.75 MPa, and the transverse normal stress of the bottom slab at the girder end is about 4.80 MPa; principal tensile stress of the top slab at the girder end is about 5.20 MPa, and the principal tensile stress of the bottom slab at the girder end is about 5.53 MPa. See Fig. 5.61 for the stress contour.

5.2.11 *Main calculation results of half diaphragm reinforcement scheme*

(1) Operating condition
Self-weight, prestress, secondary dead load and live load are taken into account as the external force of the girder body. According to the influence line for reaction, the location of live load action is the location where the maximum bearing reaction exists.

Under operating condition, the transverse normal stress of the top slab at the girder end is about 2.39 MPa, and the transverse normal stress of the top surface of the bottom slab at the girder end is about 3.70 MPa; principal tensile stress of the top slab at the girder end is about 3.43 MPa, and the principal tensile stress of the bottom slab at the girder end is about 4.57 MPa. See Fig. 5.62 for the stress contour.

(2) Structure stress under girder transport condition
Self-weight, prestress and load of girder transport vehicle are taken into account as the external force of the girder body. According to the influence line for reaction, the location of load action of girder transport vehicle is the location where the maximum bearing reaction exists.

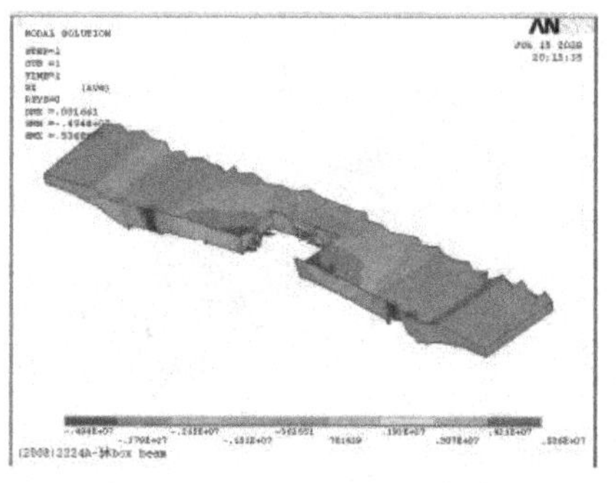

(a) Transverse normal stress of top slab

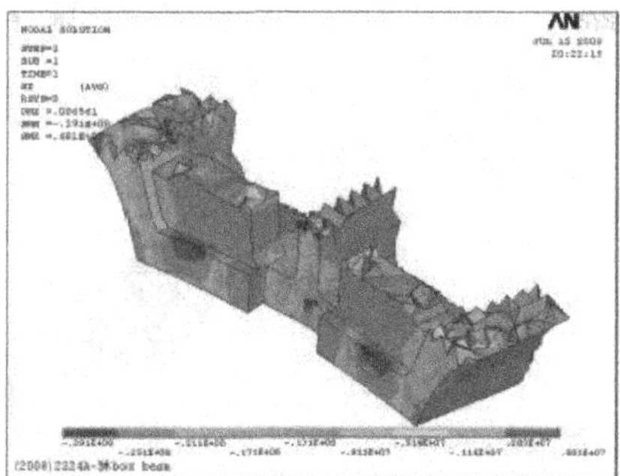

(b) Transverse normal stress of bottom slab

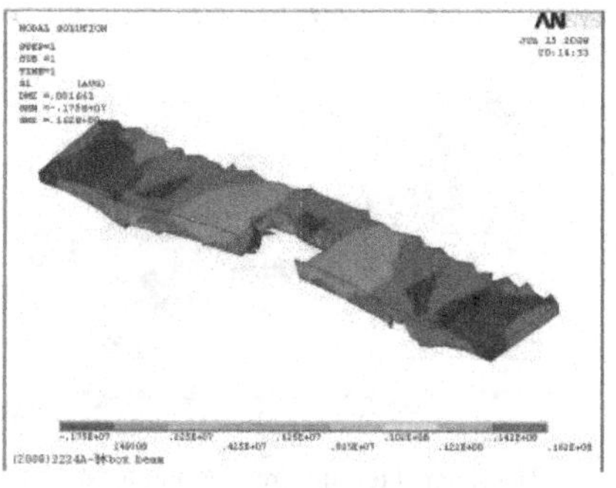

(c) Principal tensile stress of top slab

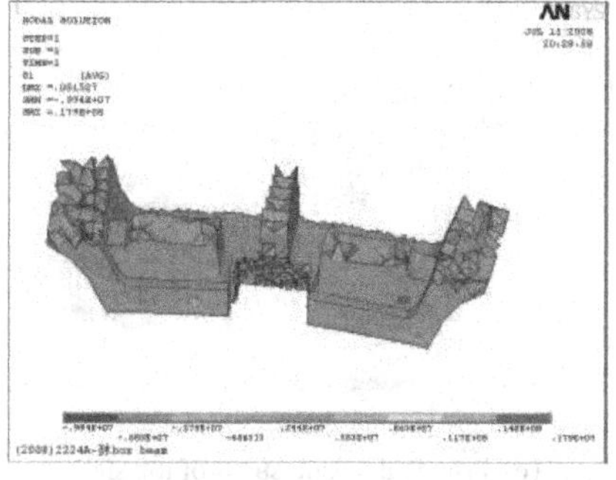

(d) Principal tensile stress of bottom slab

Fig. 5.61 Stress Contour at Girder End under Girder Erection Condition

Under girder transport condition, the transverse normal stress of the top slab at the girder end is about 2.42 MPa, and the transverse normal stress of the bottom slab at the girder end is about 3.70 MPa; principal tensile stress of the top slab at the girder end is about 2.65 MPa, and the principal tensile stress of the bottom slab at the girder end is about 4.61 MPa. See Fig. 5.63 for the stress contour.

(3) Structure stress under girder erection condition

Self-weight, prestress and load of girder erection vehicle are taken into account as the external force of the girder body. According to the influence line for reaction, the location of load action of girder erection vehicle is the location where the maximum bearing reaction exists.

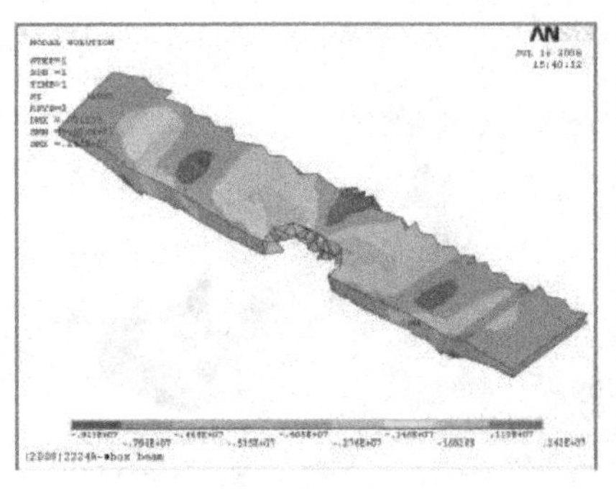

(a) Transverse normal stress of top slab

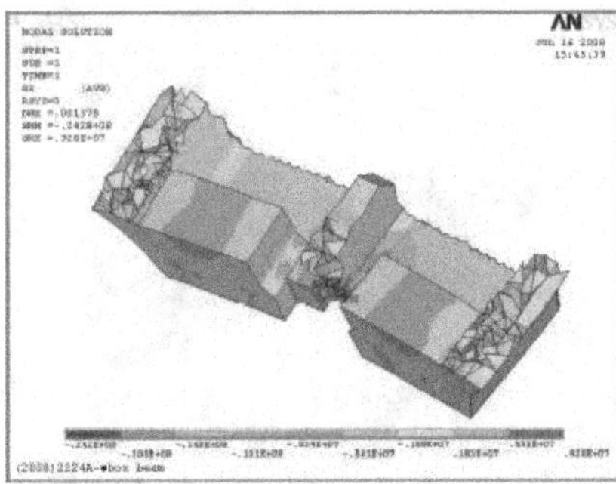

(b) Transverse normal stress of bottom slab

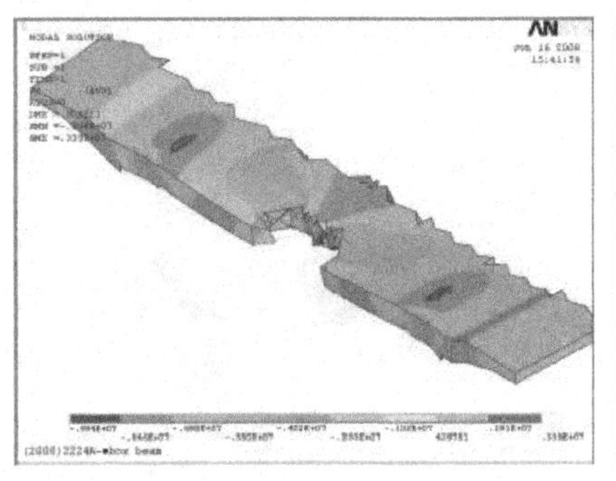

(c) Principal tensile stress of top slab

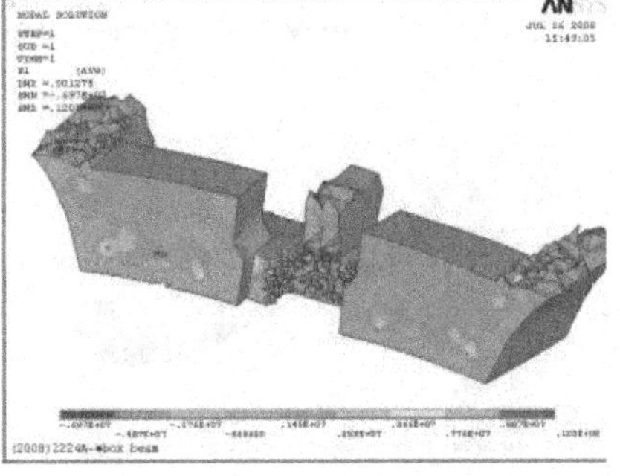

(d) Principal tensile stress of bottom slab

Fig. 5.62 Stress Contour at Girder End under Operating Condition

Under girder erection condition, the transverse normal stress of the top slab at the girder end is about 4.85 MPa, and the transverse normal stress of the bottom slab at the girder end is about 4.40 MPa; principal tensile stress of the top slab at the girder end is about 4.45 MPa, and the principal tensile stress of the bottom slab at the girder end is about 5.44 MPa. See Fig. 5.64 for the stress contour.

5.2.12 *Main conclusions*

(1) The selection of height of the diaphragm meets the space requirement for installation of lifting devices during the lifting

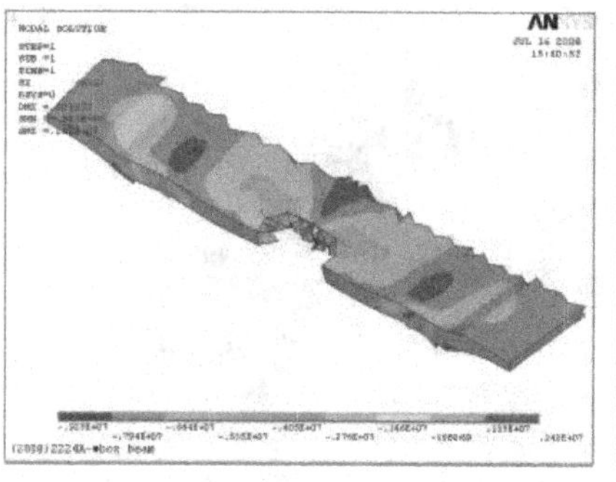

(a) Transverse normal stress of top slab

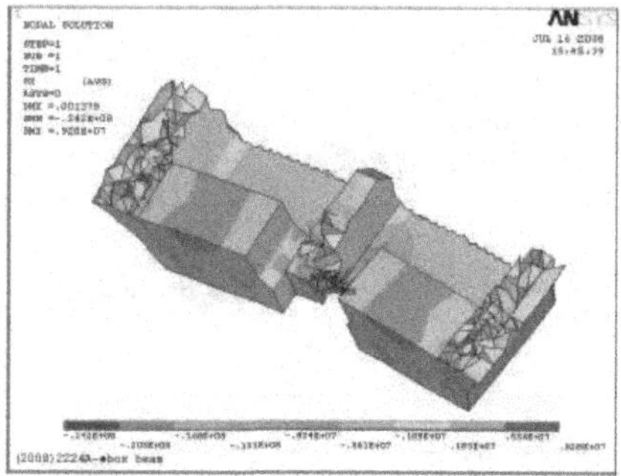

(b) Transverse normal stress of bottom slab

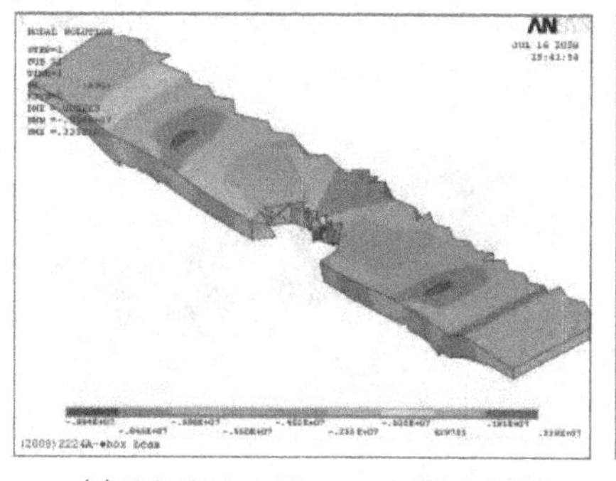

(c) Principal tensile stress of top slab

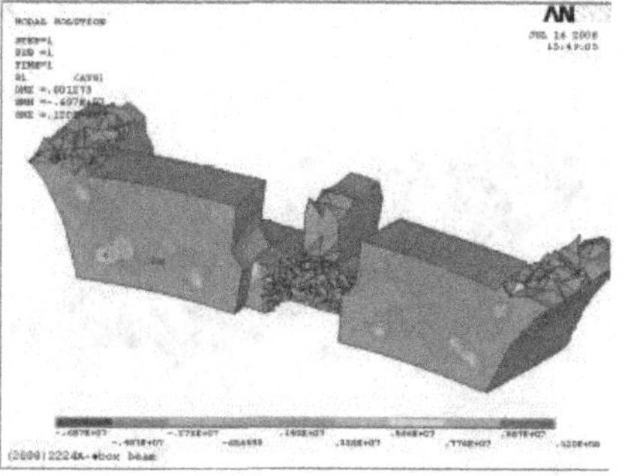

(d) Principal tensile stress of bottom slab

Fig. 5.63 Stress Contour at Girder End under Girder Transport Condition

procedure of the box girder, and the width of the diaphragm can ensure that the running channel for maintenance inside the box girder is unblocked.

(2) Both the full diaphragm scheme and the half diaphragm scheme improve, to a large extent, the stress condition of bottom slab at the bearing location at the girder end. The nominal stress under all kinds of conditions can be controlled within 5 MPa.

(3) For the prefabricated box girder, in the full diaphragm scheme, taking the impacts of shrinkage of post-poured concrete of cross wall into account, the tensile stress of a relatively large area at upper flange of bottom slab is more than 5 MPa.

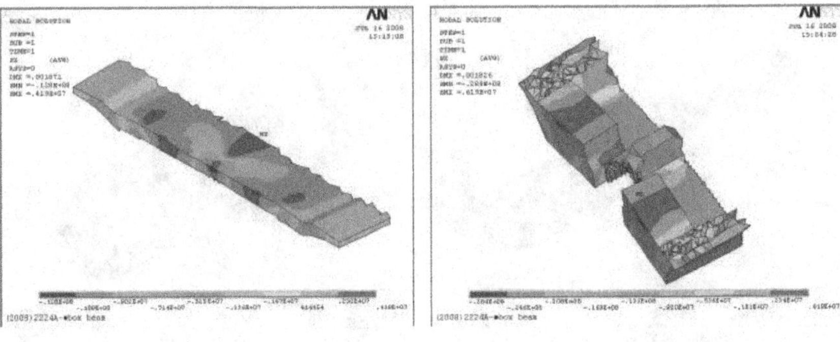

(a) Transverse normal stress of top slab (b) Transverse normal stress of bottom slab

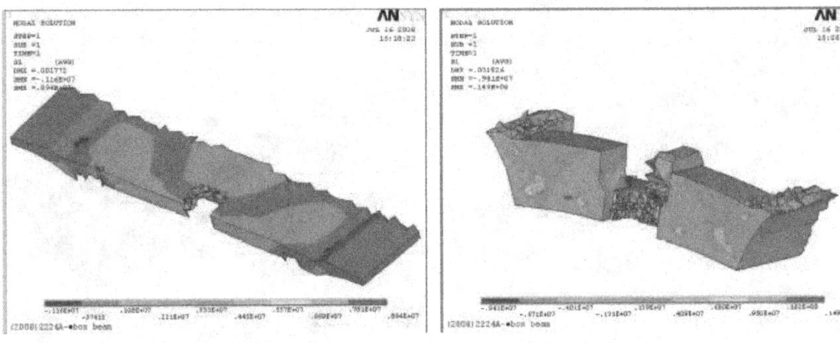

(c) Principal tensile stress of top slab (d) Principal tensile stress of bottom slab

Fig. 5.64 Stress Contour at Girder End under Girder Erection Condition

5.2.13 *Suggestions*

(1) In the scheme of setting the diaphragm at the girder end, the bar embedment scheme needs to be adopted for the prefabricated box girder. As the anchorage devices are densely distributed at the girder end, which is located in the region of high stress, the drilling and bar embedment shall avoid the anchorage devices and steel bars, and reliable anchor adhesive should be used. With the consideration of impacts of construction errors, detecting tools should be used during the construction to avoid the existing steel bars of the girder body and the prestressed pipes, so as to prevent damaging the stressed steel bars of the girder body and the prestressed anchorage.

(2) The increase of bearing size is favorable for the improvement of local stress of the girder end.
(3) As the bearing reaction at the girder end is relatively large under the girder erection condition, it can effectively reduce the stress of the girder end under the girder erection that add the bearings at the central web and the both sides, while there is still a relatively large area with tensile stress on the bottom slab. The method of girder erection can be optimized, and the locations of outriggers at both sides of central web of the girder-erecting machine can be adjusted.
(4) The pouring in the full diaphragm scheme should ensure that the concrete at the joint between diaphragm top and the top slab is dense.
(5) It is helpful to reduce the shrinkage cracks resulting from the concrete age difference by strengthening the reinforcement at the contact surface of the existing and new concrete of diaphragm and the prefabricated box girder.
(6) Compared to the design structure, although the stress level in the scheme of adding diaphragm reduces to a relatively large extent, there is still relatively large tensile stress existing under the girder erection condition. To ensure the safety of stress of the girder body and the safety of construction, each temporary bearing should be ensured to be evenly stressed and stable.

5.3 Consultation on Improvement Design of Filling Material of Xiashu Clay for the Subgrade of the High-speed Railway

5.3.1 *General*

Special consultation is carried out for the improvement method of Quaternary Upper Pleistocene clay (Xiashu clay) adopted for the subgrade filling design of a certain high-speed railway, with the fill quantity of the clay accounting for 40% of the total fill quantity.

Xiashu clay belongs to the Quaternary Upper Pleistocene (Q_3) stratum, classified as Group C or Group D filling material. Formed due to alluvial and pluvial deposition, it is distributed in the alluvial plain, high terraces and high-terrace region. The plain and the first terrace are flat and open, while the high-terrace region appears to be the mound type landscape, and the mound top is mainly perfectly round, with a gentle natural grade. The lithology is of clay, grayish yellow and brown yellow, with homogeneous texture, mainly containing clayey and silty particles, rich in iron-manganese concretions and containing calcareous concretions (sand loess-doll), and the structure is relatively compact. With developed vertical joints and prismatic structures, it belongs to the weakly expansive soil with high plasticity and high liquid limit. In the field, small gullies with herringbone slope surface are developed, forming a part with small-scale collapse, and a part with relatively high underground water level. The main physical mechanical indexes of Xiashu clay along the railway are different in the two sections.

In the first section: natural moisture content is 25.3–34.8%, void ratio is 0.73–0.99, natural unit weight is 18–20.2 kN/m^3, plastic limit is 13.8–19.5%, liquid limit is 28.3–53.8%, plasticity index is 9.0–26.4, clay content is 38%, silt content is 26%, specific area is 159.2–179.9 m^2/g, disintegration of undisturbed soil in 10 minutes to 3 hours is 65–70%, disintegration in 4–72 hours is 70–80%, residual shear strength is $c = 7-12$ kPa, $\varphi_r = 10-29°$. In Xiashu clay, the montmorillonite content is 10–18%, the free swelling ratio is 37–58%, and the cation exchange capacity is 188.17–321.58 mmol/kg, and it is of weakly $\sim$ medium expansive soil.

In the second section: natural moisture content is 10.4–22.5%, void ratio is 0.67–0.79, natural unit weight is 18.9–19.9 kN/m^3, plastic limit is 15.4–18.9%, liquid limit is 33.3–41.0%, plasticity index is 17.9–24.6, free swelling ratio is 18.5–36.0%, volume shrinkage ratio is 10.3–13.6%, disintegration of undisturbed soil in 10 minutes to 3 hours is 15–30%, disintegration in 4–72 hours is 35–60%, unconfined compressive strength under swelling limit is 8–11 kPa, residual shear strength is $c = 8-11$ kPa, $\varphi_r = 28.8-31.3°$. In Xiashu clay, the montmorillonite content is 7–10%, the cation exchange capacity is 161.24 mmol/kg, and it is of weakly expansive soil.

Comprehensively analyzed according to the geological characteristics in the field and the geotechnical test indexes, the Xiashu

Fig. 5.65 Conditions of Xiashu Clay in the Field

clay is over-consolidated and weakly ~ medium expansive soil, and it is liable to be disintegrated in case of water, with relatively poor engineering properties. Particularly, its strength and modulus will sharply decline in case of water. The Xiashu clay in the first section has a poorer engineering property due to the large content of montmorillonite in the clay minerals and large cation exchange capacity. Conditions of Xiashu clay in the field are shown in Fig. 5.65.

5.3.2 *Analysis*

As there is a small quantity of qualified filling materials in the two sections, for the Xiashu clay which cannot meet the requirements of the high-speed railway, research and analysis of improvement method, construction process of subgrade filling and method of compaction quality inspection are carried out in the design. It is considered that the requirements for the filling materials of the base course in subgrade bed of the high-speed railway can be met through improvement or strengthening of construction control. The main analysis conclusions are as follows:

(1) As a kind of silty clay or clay, the Xiashu clay is classified as Group C or Group D filling material, and it is liable to be humidified, with poor water stability and greatly differed

strength in case of water, so it cannot be directly used as the filling material of the base course in subgrade bed of the high-speed railway.

(2) After the Xiashu clay is improved by mixing the lime, the strength index is increased sharply, the compressibility is decreased significantly and the water stability is obviously improved. With good engineering property, it can be used as the filling material of the base course in subgrade bed of the high-speed railway.

(3) Corresponding to the Grade III lime, the optimal mixture amount of lime in the Xiashu clay is 5%, and the subgrade construction can be controlled as per this mixture amount.

(4) The detection data indicates that the compaction quality of the Xiashu clay improved with the mixture of lime can meet the requirements of relevant specifications.

5.3.3 *Treatment scheme*

(1) Group D fine-grained filling materials of Xiashu clay must be improved before they are used as the filling materials of embankment under the subgrade bed and the base course of the subgrade bed. Method of improvement: carry out the improvement treatment by mixing 5% lime in it; the road-mixing method should be adopted for the construction of filling materials of embankment under the subgrade bed, and the field-mixing method should be adopted for the construction of filling materials of base course of subgrade bed.

(2) Group C fine-grained filling materials should also be improved before they are used as the filling materials of embankment under the subgrade bed and the base course of the subgrade bed. Method of improvement: carry out the improvement treatment by mixing 5% lime in it; the field-mixing method should be adopted for the construction of filling materials of base course of subgrade bed, and the road-mixing method should be adopted for the construction of filling materials of embankment under the subgrade bed.

(3) When the natural moisture content of the Xiashu clay is more than 26%, it cannot be improved until it is aired to reduce the moisture content.
(4) The parameters of improved soil: the laboratory unconfined compressive strength of the base course of the subgrade bed is not less than 800 kPa, immersion saturated without disintegration within 72 hours, the strength reduction rate is less than 30–40%, and the field sampling strength is not less than 500 kPa (which is about 0.6–0.7 times of the laboratory strength). The strength index of embankment under the subgrade bed can be lowered appropriately, and the immersion saturated strength is not less than 200 kPa.

5.3.4 *Consultation conclusions*

Based on the on-site verification of the borrow area and after research of the design analysis and treatment scheme, the consulting agency thinks that the quality of the filling material of Xiashu clay is relatively poor, and there is a large difference between the research and the actual filling construction, despite that the design research proves that it can be improved by mixing the lime to meet the requirement for compaction of the filling materials. Impacted by the environment, construction process and construction equipment, in the site construction, it may be difficult for the subgrade filling to meet the requirements for compaction. It is suggested that qualified filling materials should be selected.

Xiashu clay improvement should not be adopted for the base course of the subgrade bed, and a large area of Xiashu clay improvement should not be used as the filling material of embankment under the subgrade bed. After comparison, if it is necessary to use the improved Xiashu clay as the subgrade filling material, specific requirements for the construction process, construction procedures and construction quality shall be put forward in the design document, to prevent the impacts on the construction quality of the subgrade.

According to the consultation and on-site investigation of the qualified filling materials in each worksite, in the revised design, the filling materials in the quarry is selected, instead of the borrow area of Xiashu clay, to minimize the use of improved Xiashu clay as the subgrade filling material. In total, the improved Xiashu clay to be used as the subgrade filling material in the original design is reduced by 83.3%.

5.4 Consultation on Reasonable Pier Type of the High-speed Railway

5.4.1 *General*

In the design of a certain high-speed railway, the piers with the total height $H \leq 10$ m account for 76% of the total number of piers, the piers with the total height 10 m $< H \leq 15$ m account for 18%, and the piers with the total height $H > 15$ m account for about 6%. As the high-speed railway consists of a large length of bridge, with a large number of piers, and requires large amount of materials, the selection and adoption of pier type are of great importance.

The consulting agency carries out the checking calculations of stress, strength, deformation, natural vibration frequency, stability and rigidity under the load combination of construction, girder erection and operation. The technical and economic comparison of pier bodies and foundation quantities of round-ended solid pier with top cap, rectangular solid pier and streamline round-ended solid pier is also carried out.

5.4.2 *Pier types of foreign high-speed railways*

In Germany, there are a small number of high-speed railway bridges, which are mainly viaducts and some interchanges. Taking Cologne–Frankfurt High-speed Railway as an example, with a total length of 177 km, the railway is designed with 18 bridges in total. With total length 6.06 km, accounting for 3.4% of the total length of the railway, the bridges consist of 15 viaducts, two river-crossing bridges and one interchange. As most bridges of the high-speed railway are viaducts, the bridges are mainly constructed by cast-*in-situ* at bridge sites. In Germany, the bridges are mainly designed with reinforced concrete hollow piers, with the transverse width of the pier top generally being within 6.4 m, and the longitudinal dimension generally being 2.7–4.0 m.

In France, coordination and harmony between the bridges and the surrounding cultural and natural landscapes are emphasized during the construction of high-speed railway bridges. Almost all the bridges are designed with different pier types, and each bridge is designed with a unique landscape, thus requiring the increase of relevant construction investments by about 50% at most. The high-speed railway bridges in France are mainly constructed by the incremental launching method, consisting of prestressed concrete girders and composite girders. The dimensions of the bridge piers are relatively large.

Italy is famous for the manufacturing of large-tonnage girder-erecting machine for the erection of high-speed railway bridges. The proportion of bridges to the total length of the railways in Italy is larger than that in Germany and France. Different from high-speed railway bridges in Germany and France, the bridges in Italy are mainly constructed by the prefabrication and erection method. The piers are mainly designed as solid piers with decorative grooves at both sides; cylindrical piers are adopted at the river-crossing locations.

The total length of bridges of Seoul–Busan High-speed Railway in Korea is 112 km, accounting for 27% of the total length of the railway. The girders are mainly designed as continuous girders with spans of 25 and 40 m. The continuous girders with the span of 25 m are first simply-supported and then continuously constructed, and the simply-supported part is erected by prefabrication. Different lots are designed with different types of piers, which are mainly rectangular piers and cylindrical piers. The outline dimensions of the rectangular hollow piers remain the same, with transverse width of 6 m and longitudinal length of 3.5 m. A 1 m wide, 1 m deep and 3.5 m long groove is arranged in the middle of the pier top for bearing inspection, maintenance and replacement in the operation stage.

With distinct characteristics, Japanese Shinkansen bridges take a very large proportion. Except that the interchanges and river-crossing bridges are designed as T-girders, continuous girders, continuous rigid frames and cable-stayed bridges, other bridges are designed as continuous rigid-framed bridges with small span and constructed by cast-*in-situ* at the bridge sites. For the T-girders used to cross the roads and small rivers, eight pieces of T-girders are adopted for the double-track bridges to reduce the structure height; the bridge piers are mainly designed as slab-type piers, while the cylindrical piers are adopted for the river-crossing bridges; some elevated railway stations are designed with steel tube concrete piers, to reduce the cross section size of the pier bodies.

5.4.3 *Main structural types of piers of Chinese high-speed railways*

The high-speed railway bridges in China take a large proportion. The structural types of piers used in the bridge design mainly include rectangular solid pier (Fig. 5.66(a)), round-ended solid pier (Fig. 5.66(b)), streamline round-ended solid pier (Fig. 5.66(c)), rectangular double-column pier (Fig. 5.66(d)), rectangular hollow pier (Fig. 5.66(e)), round-ended hollow pier (Fig. 5.66(f)) and cylindrical pier (Fig. 5.66(g)). Among which, the cylindrical piers are generally not adopted, except that a few cylindrical piers are adopted for the river-crossing bridges with relatively large skew angle.

The total length of high-speed railway bridges in Taiwan is 251 km, accounting for 72.8% of the total length of the railways. The bridges are mainly designed as simply supported girders with a span of 30 and 35 m, and prefabrication erection and cast-*in-situ* construction methods are adopted for the girders, among which about 55.3% girders are erected by prefabrication. Different lots are arranged with different types of bridge piers, including single rectangular column pier, single circular column pier, rectangular slab pier and pile pier. As cantilevers at bridge ends are relatively long, the longitudinal dimensions of top caps of the bridge piers are large. The transverse dimensions of the piers of high-speed railway bridges in Taiwan are relatively small.

(a) Effect Diagram of Rectangular Solid Pier

(b) Effect Diagram of Round-ended Solid Pier

(c) Effect Diagram of Streamline Round-ended Solid Pier

(d) Effect Diagram of Rectangular Double-column Pier

(e) Effect Diagram of Rectangular Hollow Pier

(f) Effect Diagram of Round-ended Hollow Pier

(g) Effect Diagram of Cylindrical Pier

Fig. 5.66 Main Structural Types of Piers of Chinese High-speed Railways

5.4.4 *Principles of pier type selection*

(1) Round-ended solid piers are generally adopted in the water section of the river-crossing bridges which has a small skew angle with the river, as well as in the flood storage areas and flood drainage areas. In the dry land section and wasteland section of the same bridge, the same pier type can be selected in a whole section to harmonize the bridge appearance.
(2) Rectangular double-column piers should be adopted for the piers with height of less than 12 m in the sections with little water or no water.
(3) Rectangular solid piers are generally arranged in the whole section of a dry land section or wasteland section.
(4) Hollow piers should be preferred if the height of most piers of the same bridge is more than 18 m.
(5) For the river-crossing bridges with relatively large skew angle, generally, single cylindrical piers should be adopted according to the flood control evaluation.

5.4.5 *Static analysis*

Static analysis calculations of round-ended solid pier, rectangular solid pier, streamline round-ended solid pier, rectangular double-column pier, round-ended hollow pier and rectangular hollow pier are carried out. 32 m + 32 m prestressed simply-supported box girder is adopted for the calculation of girder span; degree-VI seismic intensity ($a \leq 0.1\,g$) is adopted for the seismic checking calculation; the radius of curve lines of $R = 5,500$ m is adopted for checking calculation of pier body; and straight line is adopted for foundation calculation.

The geological data of the piers of a certain high-speed railway is adopted for the checking calculation of pile foundation. With the flow plastic $\sim$ hard plastic silty clay in the upper layer, and the medium dense and saturated silty soil and soft plastic silty clay in the lower layer, the geological conditions are relatively poor, as shown in Fig. 5.67.

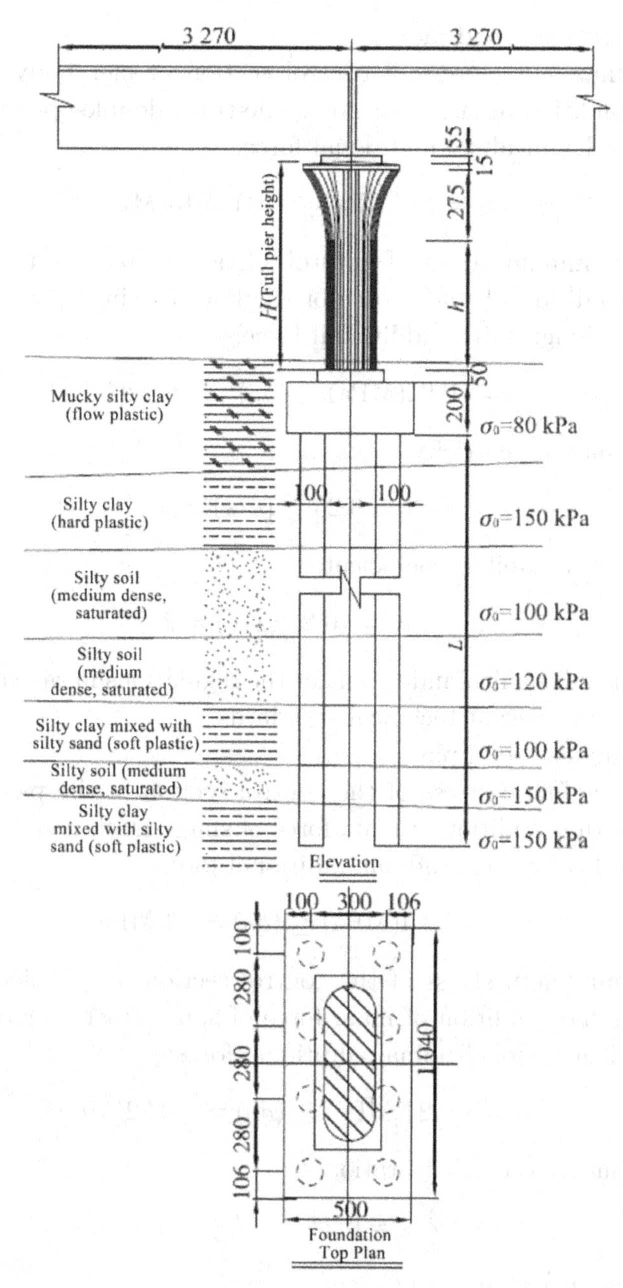

Fig. 5.67 Diagram of Checking Calculation of Pier Body and Foundation (Unit: cm)

(1) Round-ended solid pier

The maximum stress of control section of pier body is under the condition of main force of single-track double-opening heavy load + longitudinal additional force:

$$\sigma = 4.6(\text{MPa}) \leq (\sigma_0) = 13(\text{MPa})$$

The minimum stress of control section of pier body is under the condition of main force of single-track single-opening light load + longitudinal additional force:

$$\sigma = -0.02(\text{MPa}) \geq (\sigma_0) = -0.72(\text{MPa})$$

Maximum eccentricity ratio:

$$e/s = 0.23 \leq (e/s) = 0.6$$

Minimum stability coefficient:

$$K = 10.2 \geq (K) = 2$$

All the mechanical indexes meet the regulations in railway bridge code and relevant technical standards.

(2) Rectangular solid pier

The maximum stress of the control section of the pier body is under the condition of main force of single-track double-opening heavy load + longitudinal additional force:

$$\sigma = 3.4(\text{MPa}) \leq (\sigma_0) = 13(\text{MPa})$$

The minimum stress of the control section of the pier body is under the condition of main force of single-track single-opening light load + longitudinal additional force:

$$\sigma = -0.24(\text{MPa}) \geq (\sigma_0) = -0.72(\text{MPa})$$

Maximum eccentricity ratio:

$$e/s = 0.24 = (e/s) = 0.6$$

Minimum stability coefficient:

$$K = 9.7 \geq (K) = 2$$

All the mechanical indexes meet the regulations in railway bridge code and relevant technical standards.

(3) Rectangular double-column pier

The maximum stress of the control section of the pier body is under the condition of main force of single-track double-opening heavy load + longitudinal additional force:

$$\sigma = 6.5(\text{MPa}) \leq (\sigma_0) = 13(\text{MPa})$$

The minimum stress of the control section of the pier body is under the condition of main force of single-track single-opening heavy load + longitudinal additional force:

$$\sigma = 1(\text{MPa}) \geq (\sigma_0) = -0.72(\text{MPa})$$

Maximum eccentricity ratio:

$$e/s = 0.31 \leq (e/s) = 0.6$$

Minimum stability coefficient:

$$K = 15.5 \geq (K) = 2$$

Maximum compressive stress of tie beam concrete:

$$\sigma = 5.1(\text{MPa}) \leq (\sigma_0) = 13(\text{MPa})$$

Maximum tensile stress of tie beam steel bar:

$$\sigma = 140(\text{MPa}) \leq (\sigma_0) = 230(\text{MPa})$$

All the mechanical indexes meet the regulations in railway bridge code and relevant technical standards.

(4) Round-ended hollow pier

The maximum stress of control section of pier body is under the condition of main force of single-track double-opening heavy load + longitudinal additional force:

$$\sigma = 6.4(\text{MPa}) \leq (\sigma_0) = 13(\text{MPa})$$

The minimum stress of control section of pier body is under the condition of main force of single-track single-opening heavy load + longitudinal additional force:

$$\sigma = 0.98(\text{MPa}) \geq (\sigma_0) = -0.72(\text{MPa})$$

Maximum eccentricity ratio:

$$e/s = 0.29 \leq (e/s) = 0.6$$

Minimum stability coefficient:

$$K = 9.5 \geq (K) = 2$$

All the mechanical indexes meet the regulations in railway bridge code and relevant technical standards.

(5) Rectangular hollow pier

The maximum stress of control section of pier body is under the condition of main force of single-track double-opening heavy load + longitudinal additional force:

$$\sigma = 6.6(\text{MPa}) \leq (\sigma_0) = 13(\text{MPa})$$

The minimum stress of control section of pier body is under the condition of main force of single-track single-opening heavy load + longitudinal additional force:

$$\sigma = 1.1(\text{MPa}) \geq (\sigma_0) = -0.72(\text{MPa})$$

Maximum eccentricity ratio:

$$e/s = 0.3 \leq (e/s) = 0.6$$

Minimum stability coefficient:

$$K = 6.8 \geq (K) = 2$$

All the mechanical indexes meet the regulations in railway bridge code and relevant technical standards.

(6) Streamline round-ended solid pier

The maximum stress of control section of pier body is under the condition of main force of single-track double-opening heavy load + longitudinal additional force:

$$\sigma = 5.16(\text{MPa}) \leq (\sigma_0) = 13(\text{MPa})$$

The minimum stress of control section of pier body is under the condition of main force of single-track single-opening light load + longitudinal additional force:

$$\sigma = -0.56(\mathrm{MPa}) \geq (\sigma_0) = -0.72(\mathrm{MPa})$$

Maximum eccentricity ratio:

$$e/s = 0.35 \leq (e/s) = 0.6$$

Minimum stability coefficient:

$$K = 10.2 \geq (K) = 2$$

All the mechanical indexes meet the regulations in railway bridge code and relevant technical standards.

The longitudinal and transverse rigidity of the six types of piers are as shown in Fig. 5.68.

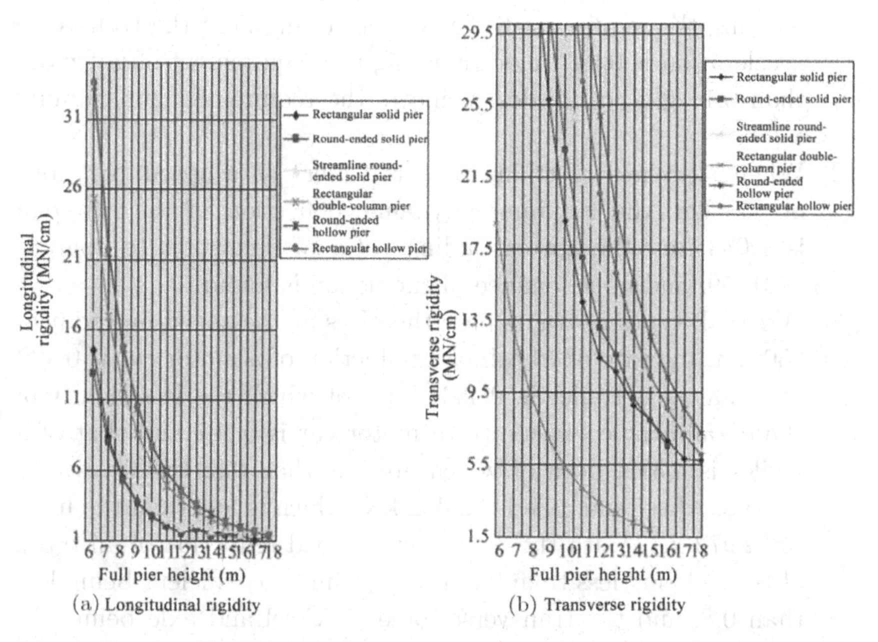

(a) Longitudinal rigidity

(b) Transverse rigidity

Fig. 5.68 Longitudinal and Transverse Rigidity of Six Types of Piers

5.4.6 *Dynamic analysis*

Dynamic analyses are carried out for rectangular piers with a height of 20 m, round-ended piers with a height of 20 and 25 m and double-column piers with a height of 12 and 15 m. Bridge discipline software is adopted for the modeling calculation of dynamic analysis, and for the train–bridge coupling dynamic analysis, MSC.PATRAN, MSC.NASTRAN and MSC.ADAMS/RAIL are adopted to establish the spatial vibration analysis model of trains and bridges, respectively for calculation. The train types include Chinese CRH, German ICE3, French TGV and Japanese 500 series train. Time domain irregularity of German low interference spectrum conversion is used as the excitation source of track irregularity.

According to the dynamic checking calculation of the double-column pier with height of 12 m:

(1) For the bridges designed as 32 m simply-supported girders coordinated with 12 m rectangular double-column piers, the maximum longitudinal vibration acceleration is 0.132 m/s^2, meeting the requirement of not more than 0.5 g; the transverse acceleration is 0.047 m/s^2, meeting the requirement of not more than 0.14 g, and the design meets the requirement for running safety.
(2) When all types of trains pass the piers at a speed per hour of 350 km, the maximum dynamic deflection in the midspan is 1.033 mm, the corresponding ratio of deflection to span is 1/31559, and no resonance phenomenon is found.
(3) When all types of trains pass the piers at the speed per hour of 350 km, the rate of wheel load reduction of a motor car is 0.422 and that of a trailer is 0.444, both of which are less than 0.6; the derailment coefficient of a motor car is 0.307 and that of a trailer is 0.312, both of which are less than 0.8; the transverse force of wheel and axle is 32.382 kN, which is less than the limit of 52.972 kN. With the rate of wheel load reduction of all types of trains being less than 0.6, the derailment coefficient being less than 0.8, and the transverse force of wheel and axle being less than the limit, the requirement for running safety is met.

(4) When the German ICE3 train passes the piers at a speed per hour of 350 km, the transverse comfort of the motor cars and trailers is good, and the vertical comfort is excellent. The acceleration of car body is qualified.
(5) When the French TGV train passes the piers at a speed per hour of 350 km, the transverse comfort of the motor cars and trailers is excellent, and the vertical comfort is good. The acceleration of car body is qualified.
(6) When the Japanese 500 series train passes the piers at a speed per hour of 350 km, the transverse comfort of the motor cars and trailers is good, and the vertical comfort is good. The acceleration of car body is qualified.
(7) When the Chinese CRH high-speed train passes the piers at a speed per hour of 350 km, the transverse comfort of the motor cars and trailers is good, and the vertical comfort is good. The acceleration of car body is qualified.

Therefore, the conclusions of the preliminary analysis are:

(1) When the height of pier body is less than 12 m, the transverse dynamic displacement of the double-column pier is similar to that of the hollow pier, while the transverse vibration acceleration of the double-column pier is larger than that of the hollow pier with corresponding height.
(2) When the height of pier body is within 12–15 m, the transverse dynamic displacement of pier top of the double-column pier is larger than that of the hollow pier with corresponding height; besides, the transverse vibration acceleration of the double-column pier is larger than that of the hollow pier with corresponding height; the responses of the rolling stock on the double-column pier are poorer than that on the hollow pier.

5.4.7 *Analysis of work quantities and construction costs of bridge piers and foundations*

The piers with the same support of 32 m simply-supported box girder, with same pier height and same geological conditions are adopted for the analysis calculation as per the same technical standards and specifications to determine the work quantities of pier body and foundation.

According to the *Code for Design on Subsoil and Foundation of Railway Bridge and Culvert*, the bearing platform shall meet the requirement for rigid angle, and structural calculation and analysis comparison of the bearing platform shall be carried out as per different calculation theories, so as to arrange reasonable quantities of steel bars.

The boundary dimensions of the double-column pier bottom and the pier bottom formed between the two columns are relatively large. When the 8 or 10 pile foundations are arranged, the thickness of bearing platform of 2 m can meet the requirement for rigid angle. As the dimensions of pier bottom of the solid pier are smaller than the dimensions of double-column pier bottom, the bearing platform with a thickness of 2 m cannot meet the requirement for rigid angle, and the bearing platform should be heightened.

The analysis of construction costs of rectangular solid pier, round-ended solid pier, rectangular double-column pier, rectangular hollow pier, round-ended hollow pier and their foundations is as shown in Fig. 5.69.

The following conclusions can be obtained according to the comprehensive costs of masonry and steel bars:

(1) When the pier height is more than 10 m, the investment on the construction of round-ended solid pier is close to that of the rectangular solid pier, and the investment on the round-ended hollow pier is close to that of the rectangular hollow pier; however, the hollow pier is more economical than the solid pier.
(2) When the pier height is less than 10 m, it is more economical to be designed as a round-ended solid pier rather than a rectangular solid pier.
(3) When the pier height is less than 12 m, the double-column pier is the most economical design compared with other four pier types.
(4) When the pier height is more than 12 m, the double-column pier is more economical than the two types of solid piers, but it has less economical efficiency than the hollow pier.

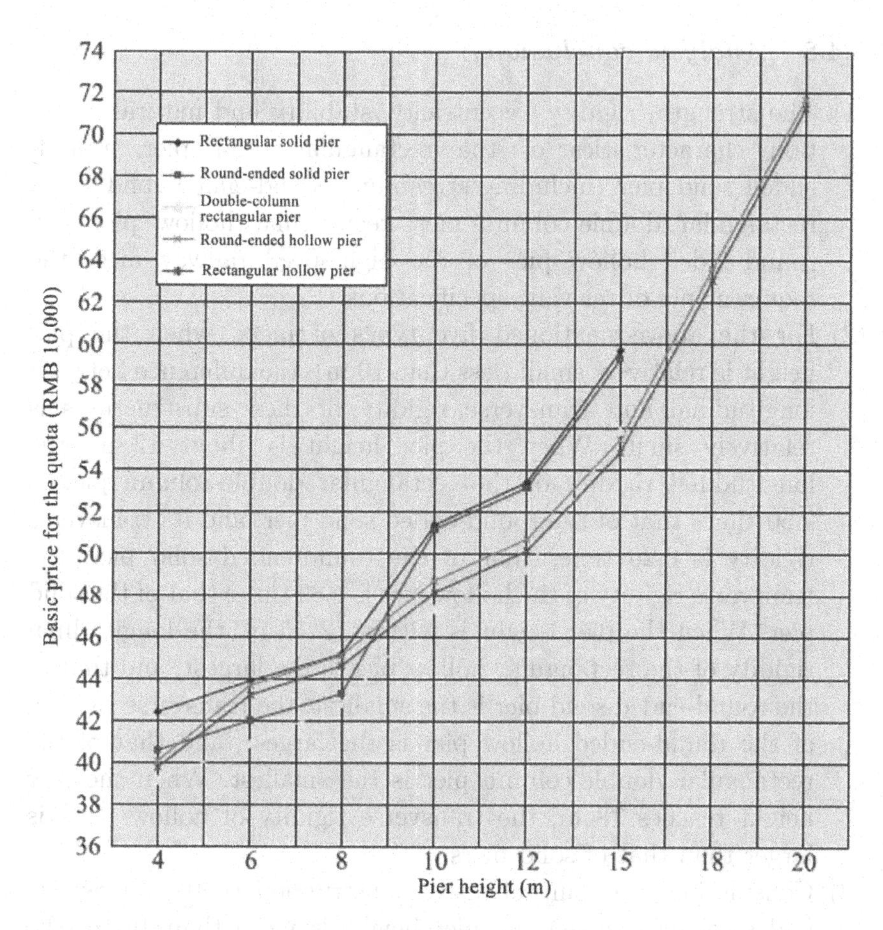

Fig. 5.69 Investment Analysis Curve of Pier and Foundation

In addition, the longitudinal and transverse rigidity of the streamline round-ended solid pier are basically the same with those of the traditional round-ended solid pier.

The longitudinal and transverse rigidity of the streamline round-ended solid pier are consistent with those of the traditional round-ended solid pier, while the quantity of concrete and steel bars used for the streamline round-ended solid pier is less than that for the traditional round-ended solid pier. Seen from the effect diagram, the transition between pier body and the top cap of the streamline round-ended solid pier is smoother, with more coordinated streamline with the girder.

5.4.8 *Analysis conclusions*

(1) The strength, rigidity, eccentricity, stability and natural vibration characteristics of the rectangular solid pier, round-ended solid pier (including streamline round-ended solid pier), rectangular double-column pier, rectangular hollow pier and round-ended hollow pier of the high-speed railway meet the requirements of relevant specifications.

(2) For the above-mentioned five types of piers, when the pier height is relatively small (less than 10 m), the difference between longitudinal and transverse rigidity of their substructures is relatively small. When the pier height is about 12 m, the longitudinal rigidity of the rectangular double-column pier is 2.50 times that of the round-ended solid pier, and its transverse rigidity is 0.26 times that of the round-ended solid pier; the transverse rigidity of the hollow pier is 1.89 times that of the solid pier. When the pier height is within 12–15 m, the longitudinal rigidity of the rectangular hollow pier is the largest, and that of the round-ended solid pier is the smallest; the transverse rigidity of the round-ended hollow pier is the largest, and that of the rectangular double-column pier is the smallest. When the pier height reaches 18 m, the transverse rigidity of hollow piers is larger than that of solid piers.

(3) Considering the comprehensive construction costs of masonry and steel bars, when the pier height is more than 10 m, the investment on construction of round-ended solid pier is close to that of the rectangular solid pier, and the construction cost of round-ended hollow pier is close to that of the rectangular hollow pier, while the construction cost of hollow piers is less than that of solid piers. When the pier height is less than 10 m, the construction cost of the round-ended solid pier is less than that of the rectangular solid pier. When the pier height is less than 12 m, the construction cost of the double-column pier is the least compared with the other four pier types. When the pier height is more than 12 m, the construction cost of double-column pier is less than the solid pier, but it has less economical efficiency than the hollow pier.

(4) Compared with the piers with top caps, the bearing platforms of the piers without top cap can meet the requirement for rigid angle, with less masonry work and reduced excavation work of the bearing platform.

5.5 Consultation on the River-crossing Tunnel Works of a Certain High-speed Railway

5.5.1 *General*

The total length of the scope of the river-crossing tunnel works of a certain high-speed railway is 10.8 km. According to the geological survey and design, the shield tunnel passes through the bedrock, half-rock and half-soil and Quaternary overburden strata. See Fig. 5.70 for the longitudinal cross section of the tunnel. Except that open cut is adopted for the approach sections, the open cut and buried sections and the working shafts at the entrance and exit, shield method is adopted for the tunnel construction.

5.5.2 *Engineering geology*

(1) Topography and landform
The tunnel is located in the alluvial plain of the delta, where the terrain is flat and open. Villages, buildings, ponds, roads and rivers and ditches are densely distributed in the area, and the water system is developed. The line passes the river channel, as well as many small rivers, and the river channel is characterized by deep water and fast flow. See Fig. 5.71 for the landforms at the tunnel entrance and the working shaft.

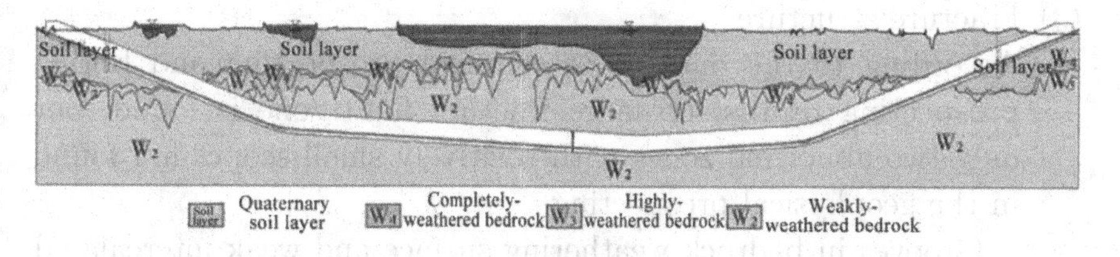

Fig. 5.70 Diagram of Tunnel Cross Section

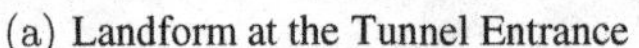

(a) Landform at the Tunnel Entrance

(b) Landform at the Working Shaft of Entrance

(c) Landform at the Tunnel Exit

(d) Landform at the Working Shaft of Exit

Fig. 5.71 Landforms at the Tunnel Entrance and Exit as Well as the Working Shafts

(2) Stratum

According to the lithology, the bedrock mainly consists of mud rock, pelitic siltstone and siltstone and fine sandstone in some local parts. According to the degree of weathering, the bedrock is classified as completely weathered layer, highly weathered layer and weakly weathered layer.

(3) Fracture structure

According to the data of geophysical prospecting and boring prospecting results, no large fracture fault zone is found, but only two abnormal zones with relatively small scopes are found in the geophysical prospecting.

Grooves in bedrock weathering surface and weak intercalated layers are found in some section in the river; water leakage is found in some drilling holes during the construction, indicating that the cracks are relatively developed, and may come from the fracture effects.

5.5.3 *Shield tunnel*

The inner diameter of the shield tunnel is 9.8 m, and the segment is 500 mm thick and 2.0 m wide. The segmentation mode of "7+1" is adopted; the strength grade of concrete is C50, and the impervious grade is S12.

5.5.4 *Engineering characteristics*

The entrance section of the tunnel passes enters the river channel after passing through two small rivers. Air-cushion-type slurry balanced shield with large diameter is adopted for the construction in the soil and rock strata with non-homogeneous hardness, and the construction scheme of underwater "construction in opposite direction, butt joint underground and disassembly in the tunnel" is adopted for the shield.

5.5.5 *Main technical problems of the work*

(1) Selection of shielding machine in the composite stratum with large diameter, long distance and high water pressure
It is the most important and difficult thing in the tunnel work to select an appropriate type of shielding machine to meet the requirement of shield tunneling in the composite stratum with large diameter, long distance and high water pressure.

(2) Slurry balance shielding machine with large diameter passing through the complex stratum in a large distance
The shield tunnel of the tunnel work is constructed by long-distance advancing at the river bottom. The inner diameter of the circular tunnel is 9.8 m, and the single end shield is advanced for about 5,000 m. The requirements for the operation and maintenance of the whole system are higher.

The shield tunneling needs to pass through several strata including silty clay, mucky clay, fine sand, medium sand, coarse sand, completely weathered ~ weakly weathered argillaceous sandstone, siltstone and fine sandstone; in the bedrock section, it also needs to pass through the fault or joint concentrated zone. The requirements for tool adaptation, slurry system and control of parameters in the tunneling are higher.

(3) Wear and replacement of tool
In the rapid construction of shield, it is key to reduce the tool wear and the tool replacement. Besides, tool replacement under high water pressure is of large safety risk and difficulty.

(4) Butt joint in the river
Butt joint and disassembly of the shielding machine are carried out in the river, so there are great construction risks. The reinforcement and water stop of the butt joint section and the support of the shield after the disassembly of shielding machine should be perfectly safe.

(5) Construction of connecting passage
In total, 11 connecting passages are arranged in the shield section at the tunnel entrance. Some connecting passages are located in the mucky soil and fine siltstone strata. Special grouting and construction in freezing method should be carried out for the excavation in case of rich underground water, and construction difficulty and risks are relatively large.

(6) Construction with shielding machine in shallowly-buried sections and adjacent sections
In the sections passed by the shielding machine, the minimum depth of the covering soil is only 5 m; the lowest location in the river channel is only about 8 m away from the tunnel top; the distance between the up and down tracks in the launching section is only 0.5 times of the tunnel diameter; under these conditions, it is difficult to control the axis control, ground settlement and parameter stability.

5.5.6 *Construction conditions on the site*

Four sets of φ11 182 mm slurry shielding machines are used to carry out the tunnel construction from the ends of the two tunnels in the opposite direction, respectively. The construction scheme of "underwater construction in opposite direction, butt joint underground and disassembly in the tunnel" is adopted.

Generally, the construction quality of the shield tunnel is good. However, there are still some problems in the aspects of construction progress, overall floating of some tunnel portal sections and the assembly quality of the tunnel (segment dislocation, damage of edges and corners of segments and water leakage and seepage at some part of the segments).

(1) Construction progress
The construction progress of the shield tunnel is not consistent with the design construction progress index.

(2) Construction quality of the shield tunnel
According to the field investigation and survey, the quality of on-site assembly of the shield segments is good. However, there are still such problems as overall floating of some tunnel portal sections, dislocation of some segments of the shield tunnel, damage of edges and corners, and local water leakage.

(3) Construction quality of open cut sections and shield shafts
In the tunnel entrance section, there are still about 100 m to be open cut; in the exit section, except for the phase II plugging of the shield shaft, the construction of open cut and shield shaft is completed. There is no safety quality accident during the construction of open cut section and shield shaft, and the construction quality is generally good.

5.5.7 *Relevant countermeasures and suggestions*

Consultation suggestions on unsatisfied tunnel construction progress, overall floating of some tunnel portal sections and severe tool wear mainly include:

(1) Shielding machine
It is appropriate to select the slurry balanced shielding machine for the tunnel work, with consideration from the aspects of stratum permeability (as shown in Fig. 5.72, the permeability coefficient for the shield passing through the stratum is $10^{-3} - 10^{-5}$ m/s), stratum particle grade (as shown in Fig. 5.73,

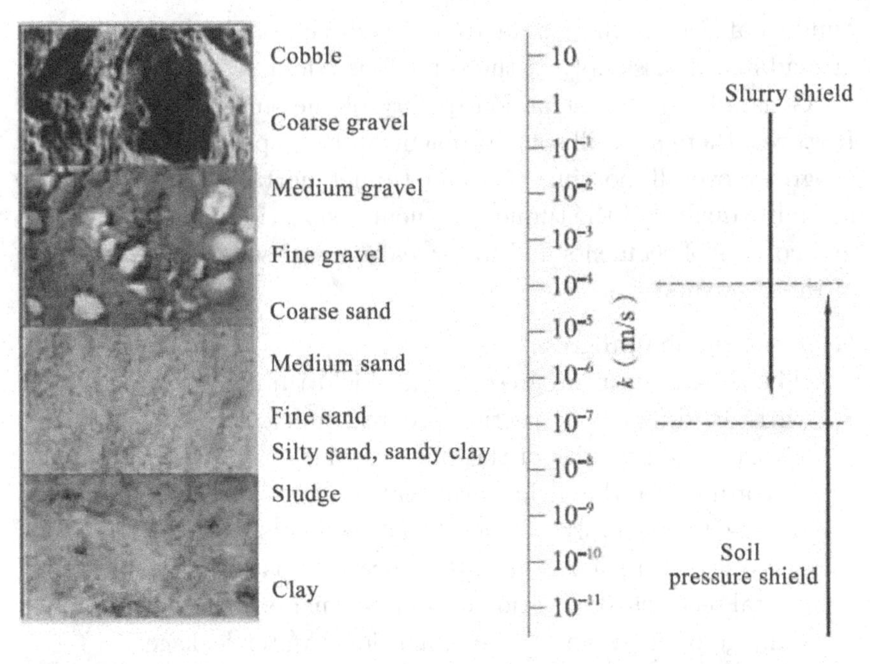

Fig. 5.72 Diagram of Relationship between Stratum Permeability and Shield Type Selection

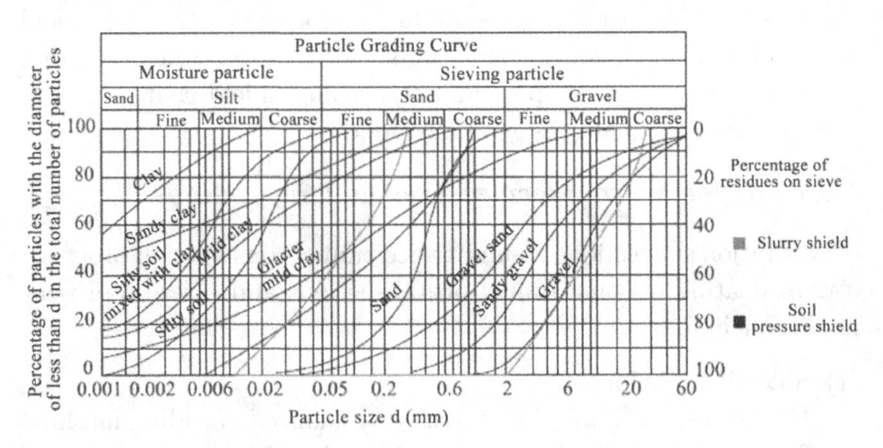

Fig. 5.73 Diagram of Relationship between Stratum Particle Grade and Shield Type Selection

according to the analysis results of stratum particles passed by the shield, the particles with diameter more than 0.075d account for more than 80%) and head pressure (the maximum head pressure of the tunnel is 0.67 MPa; when the head pressure is more than 0.3 MPa, the soil pressure balance shield is not applicable). However, in some aspects, the consulting agency thinks that corresponding optimizations can be made.

(2) Form of cutter head

Most parts of the tunnel pass through the rock stratum of pelitic siltstone, siltstone and fine sandstone, with the soft soil layer at the upper part consisting of mucky soil, silty clay, silty fine sand and medium-coarse sand, and the rock layer at the lower part consisting of pelitic siltstone, siltstone and fine sandstone. Therefore, the cutter head shall be designed mainly for the purpose of smashing the hard rocks, and also with the consideration that there is the soft soil stratum such as sand layer in the starting section. The targeted design shall be carried out (the hobbing cutter can be replaced by the scraper and slicer in the soft soil or sand layer). To effectively reduce the tool wear, the open ration of the cutter head can be enlarged appropriately.

Although the cutter head of the tunnel shielding machine is designed based on the above ideas, there are still some problems:

(i) The number of hobbing cutters (which can be replaced by the teeth cutter in the sections with soft soil stratum) is relatively small, with single form (which is mainly single blade, with relatively sectional area).

(ii) Although the open ratio of the cutter head reaches 37%, the number and sizes of the central opens are relatively large (the open at the cutter head center takes a large proportion of the cutter head area), and the open ratio at the surrounding of the panel is relatively small. To a certain extent, the removal of residual soil at the surrounding of the shield panel is affected, and it is easy for the cutter head to be incrusted.

(iii) The surrounding of the cutter head should be of circular arc shape, and be arranged with a certain number of hobbing cutters (which can be mutually replaced with the teeth cutter).

(iv) All types of tools should be reasonably combined and matched, and the combined height of the hobbing cutter (or teeth cutter) and the scraper can be further optimized (the combined height difference between the teeth cutter and the slicer in soft and weak strata should be within 30–60 mm, and the combined height difference between the hobbing cutter and the scraper should be within 90–130 mm). To reduce the tool wear, the residual soil (removed by muddy water) shall be mainly "smashed for removal." It can be seen from Fig. 5.74 that for the two different types of residual soil, the tool wear in the block and granulous residual soil has a lighter degree than that in the sandy and silty residual soil.

(v) To ensure that the shielding machine can successfully complete the long-distance tunneling task, it is suggested that at the key positions (such as the cutter head panel,

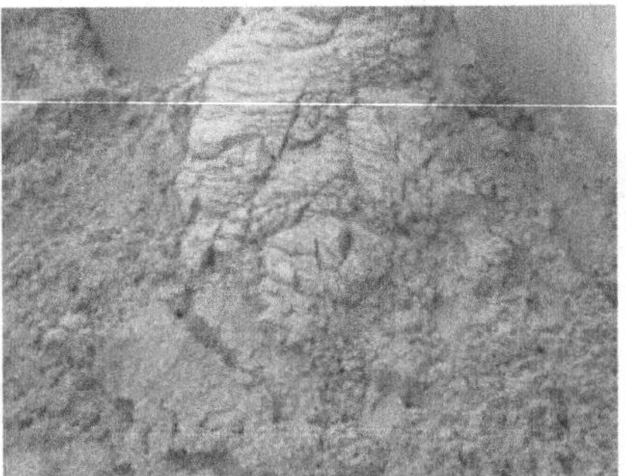

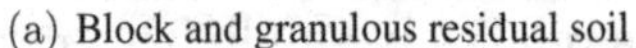

(a) Block and granulous residual soil

(b) Sandy and silty residual soil

Fig. 5.74 Two Different Types of Residual Soil

main bearing and mud discharge pipe) of the shielding machine, the wear-resistant materials should be overlaid in the manner of hardening.

(3) Cutter

Based on the shield construction principles, teeth cutter and scraper should be mainly adopted in the sections with soft and weak strata, and hobbing cutter should be mainly adopted in the sections with hard rock. After the cutter head of the shielding machine is determined, it is suggested that detailed research should be carried out in the following aspects:

(i) Appropriateness of mutual replacement of hobbing cutter and teeth cutter under different stratum conditions should be determined. According to the research on the working mechanisms of all types of cutters and the damage conditions, the appropriate cutter including teeth cutter, hobbing cutter, slicer and scraper should be selected for different strata.

(ii) The design of all types of cutters should be optimized. For example, increase the sectional area of the hobbing cutter edge; change the single type hobbing cutter face into multi types of faces inset with the wear-resistant materials; research the possibility of providing the wear-resistant materials for different types of cutters.

(iii) The arrangement of cutter head and cutters should be optimized. The arrangement of cutter head and slicer, scraper, profiling cutter and hobbing cutter should be further optimized through the research on the construction path and excavation mechanism of the cutter head and cutters.

(iv) According to the research on all types of wear-resistant materials, select appropriate additional material for the wearing layer of such key positions of the shield as cutter, cutter head and sludge discharge pipe.

(4) Additive
The construction contractor has researched several types of additives (such as water, foam and bentonite, etc.). It is suggested that research on single addictive and the combination conditions of these additives should be carried out. Besides, tentative research on other types of additives (such as foaming agent and all types of chemical additives) which are not used should also be carried out to find out the appropriate additives capable of improving the construction progress and reducing the cutter wear.

(5) Overall tunnel quality control in the tunnel portal section
Three shielding machines have been successfully started on the site. According to the construction conditions on the site, there are still some deficiencies in the quality control of the tunnel portal section after the starting:
(i) Overall floating of the shield tunnel is found.
(ii) Compared to the assembly quality of other sections, relatively severe dislocation is found.

In order to ensure the high-quality completion of shield tunnel construction in the tunnel portal section by the fourth shielding machine, the following suggestions are put forward on the construction of shield tunnel in the tunnel portal section:

(1) In the tunnel portal section, the covering soil is relatively shallow and permeable; the shield tail is lighter than the shield head and there is a tendency of relative "floating" during the construction, so the tendency of "pitching" of the shield can be seen directly. Under this condition, it is difficult to ensure the assembly quality of the shield segments, and it is the fundamental cause of dislocation of shield segments. In this aspect, to prevent the "pitching" during the construction of shield tunnel portal section, it is suggested that the counterweight should be provided for the shield tail.

(2) In the synchronous grouting, the consistency, setting time and strength of the grouting slurry are key factors. The synchronous grouting materials with large consistency, short initial setting

time and large strength should be adopted for the shield segments in the water-saturated stratum.

(3) According to the metro construction experience, to ensure the construction quality of the longitudinal axis of the shield tunnel, during the construction, the longitudinal construction axis should be 10–20 mm lower than the theoretical one.

(4) The control of shield tunneling parameters is one of the key technologies during the shield construction. Even under the same engineering hydrogeological conditions, the tunneling parameters of the same shielding machine controlled by different shield operators are different. Therefore, it is necessary to summarize the tunneling parameters of the shielding machine in all types of strata on the site. The shield tunneling parameters applicable to the hydrogeological conditions in this lot and all the specific conditions on the site should be found out through constant summaries, so as to improve the shield tunneling speed and the assembly quality of the segments.

(5) According to the actual site conditions, further strengthen the management, standardize the construction requirements, further improve the shield construction quality and minimize such deficiencies as segment dislocation, damage of segment edges and corners, and local water leakage of the segments.

(6) Connecting passage and butt joint in the river
As the connecting passage and butt joint in river have not yet been implemented, the relevant suggestions will be put forward based on the actual conditions on the site.

(7) There is no safety quality accident during the construction of open cut section and shield shaft, and the construction quality is generally good. At this stage, leaking stoppage and repair are demanded in a few positions with water leakage. During the construction of leaking stoppage, appropriate materials and technologies should be selected and the original integrity of the structural concrete and steel bars should be ensured.

(8) Strictly follow the construction safety disciplines to ensure the production safety. Special attention shall be paid to the compliance of the requirements and regulations of relevant laws, regulations, specifications, rules and regulatory documents, and five signboards (signboards for the project summary, the list of management personnel and complaints hotline, the fire control and safeguard, the work safety, and the civilized construction and environmental protection) and one map (layout plan of construction site) should be set at a prominent place at the site entrance. Adequate protection facilities should be set. All construction personnel must wear safety helmets, work clothes and work shoes for production safety, and the operators in the tunnel are strictly prohibited to drink or smoke. Personnel of special types of work must be trained and obtain the relevant certificates before assuming their roles. All workers shall be provided with the "three-level" education before they enter the site, and be assessed and issued with work certificates. All the mechanical equipment shall be provided with protective covers for safe operation and detailed safety operating points. All the vertical and horizontal transport machinery such as hoisting frame should be provided with enclosures, and alarm lights and alarm bells should be arranged if necessary. Before construction, safety precautions for the accidents must be taken based on careful research to prevent potential risks in the construction through the shield method under specific geological and operational conditions. Special attention must be paid to the prevention of potential unfavorable conditions during the operations under pressure. Emergency countermeasures and measures for the emergency incidents must be formulated and implemented in advance, and the well-equipped temporary first-aid station and medical care personnel should be provided. The material depots, substations, ventilation facilities and all the temporary facilities in the construction site should be arranged with lightning protection facilities, and the ground resistance should

be checked regularly to prevent the lightning stroke. Round-the-clock safety lighting and necessary alarm lights should be provided for the main operation places and temporary safety passage for evacuation, so as to prevent all kinds of potential accidents.

References

[English] Bryan Magee and Ji Guibao (Interpreter). *The Story of Philosophy.* Beijing: SDX Joint Publishing Company, 2009.

Chen Guang. *Philosophy of Science and Technology-Theory and Methods.* Chengdu: Southwest Jiaotong University Press, 2003.

Chen Lie. Review on Significant Technical Issues in Design and Consultation on Wuhan-Guangzhou Railway Passenger Dedicated Line. *Railway Standard Design*, 2010 (1): 6–9.

Chen Lie and Xu Youding. Main Technical Characteristics of Design on Shaoguan-Huadu Section of Wuhan-Guangzhou Railway Passenger Dedicated Line. *Railway Standard Design*, 2010 (1): 19–22.

Chen Lie, Li Xiaozhen, Liu Dejun and He Gengyu. *Composite Deck System of Through Tied-Arch Bridge on High-Speed Railway. China Railway Science*, 2007 (5): 37–42.

China Railway Eryuan Engineering Group Co., Ltd. Guiding the Consultation Work of High-speed Passenger Dedicated Railways under the Scientific Outlook on Development. *Proceedings of Summit Forum for Studying and Practicing the Scientific Outlook on Development in the Engineering Consultation Industry.* Beijing, 2008: 110–125.

China Railway Eryuan Survey and Design Group Co., Ltd. *A Complete Package for Building High Capacity Nanjing–Kunming Main Line in Complicated Geographical Mountainous Areas.* Chengdu, University of Electronic Science and Technology of China Press, 2000.

Editorial Committee of *History of China Railway Bridges. History of China Railway Bridges.* Beijing: China Railway Publishing House, 1987.

Hua Maokun. *Speed-up History of China Railways.* Beijing: China Railway Publishing House, 2002.

He Huawu. *Proceedings of Railway Engineering Technology.* Beijing: China Railway Publishing House, 2007.

Hu Yifeng and Li Nufang. *Design Theory of Ballastless Track-Subgrade for High Speed Railways*. Beijing: China Railway Publishing House, 2010.

Jiang Cheng, Wang Jijun, Hu Suoting and Jiang Ziqing. Structure and Key Technology of Ballastless Track on Passenger Dedicated Line. *Journal of Railway Engineering Society*, 2008 (supplement): 168–181.

Li Shibin. The Development Trends of High-speed Railway Technology in the World. *World Railway*, 2007 (6): 48–49.

Liu Jifeng, Dong Yeqing and Wu Jianhong. The High-Speed Railway in the World Stepping into Full Development Times-Interview Mr. Ignacio Bahon the Director of International Union of Railways (UIC), High-Speed Department. *World Railway*, 2007 (10): 16.

Li Dianlong. Enlightment for the International Consultation on the Design of the Dedicated-Passenger Railway Line. *Railway Standard Design*, 2005 (10): 1–3.

Li Yinghong, Zhu Dong and Wang Ruiyuan. Engineering Consultation Mode and Consultation Organization for Wuhan-Guangzhou Passenger Dedicated Line Wuhan-Guangzhou Passenger Dedicated Line Co., Ltd. *Collection of Wuhan-Guangzhou Passenger Dedicated Line Construction Technology* (Episode 1). Chengdu: Southwest Jiaotong University Press, 2007: 468–471.

Liu Hui, Zhu Ying and Chen Lie. Review and Thoughts on the Construction of Passenger Dedicated Line (High Speed Railway) in China. *Journal of Railway Engineering Society*, 2008 (supplement): 21–31.

Mao Junjie. *The New Line Project of Cologne-Frankfurt High-speed Railway in Germany*. Scientific and Technical Information Research Institute, Ministry of Railways, 2004.

Qian Lixin. *High-speed Railway Technology in the World*. Beijing: China Railway Publishing House, 2003.

Scientific and Technical Information Research Institute, Ministry of Railways. The Technology and Equipment of Nuremberg-Ingolstadt New Line in Germany. *World Railway Trend*, 2007 (22): 19.

Scientific and Technical Information Research Institute, Ministry of Railways. *Study of Current Situations and Development Trends of High-speed Railway* 2003, 6.

Scientific and Technical Information Research Institute, Ministry of Railways. Establishment of American High-speed Railway Association. *World Railway Trend*, 2010 (2): 28–29.

Scientific and Technical Information Research Institute, Ministry of Railways. High-speed Test of French TGV. *World Railway Trend*, 2008, 3(2): 4.

The Committee of Technology Summary of Chengdu–Kunming Railway. *Chengdu–Kunming Railway*. Beijing: China Railway Publishing House, 1980.

TB10504-2007. *Preparation Methods for Pre-Feasibility Study, Feasibility Study and Design Documents of Railway Construction Project*. Beijing: China Railway Publishing House, 2007.

Wang Xiaohong and Tan Kehu. *Some Thoughts on the Foundation of Developing High-speed Railway. Railway Economic Research*, 2009 (4): 9–15.

Wang Qichang. *Civil Engineering of High-speed Railways.* Chengdu: Southwest Jiaotong University Press, 1999.

Wu Kejian. Development and Innovation of Engineering Technology for Passenger Dedicated Line. *Journal of Railway Engineering Society*, 2008 (supplement): 1–10.

Wu Mingyou and Mi Long. The Development and Innovation of Technology Standard for Passenger Dedicated Line. *Journal of Railway Engineering Society*, 2008 (supplement): 11–20.

Xie Xianliang. *Japan High-speed Railway Technology.* Scientific and Technical Information Research Institute, Ministry of Railways, 2005.

Zhang Jianjing, Chen Lie and Han Pengfei. *Analysis on the Destruction Features of Earthquake Bridges and Anti-seismic Action of Damping Bearing. Analysis and Investigation on Seismic Damages of Projects Subjected to Wenchuan Earthquake.* Beijing: Science Press, 2009 (4): 643–650.

Zhang Jingbo. The Development History of World Light Railway. *World Railway*, 2007 (11): 46–48.

Zhao Guotang. *Ballastless Track Structure of High-speed Railways.* Beijing: China Railway Publishing House, 2006.

Zheng Jian. *High-speed Railway Bridges of China.* Beijing: High Education Press, 2008.

Zhi Ying and Chen Lie. Technical Strategies for the Design and Consulting of Wuhan-Guangzhou Passenger Dedicated Line. Wuhan-Guangzhou Passenger Dedicated Line Co., Ltd. *Collection of Wuhan-Guangzhou Passenger Dedicated Line Construction Technology* (Episode 1). Chengdu: Southwest Jiaotong University Press, 2007: 432–442.

Zhu Ying and Yi Sirong. *Theory and Methods of Dynamic Analysis on Curve Parameters of High Speed Railways.* Beijing: China Railway Publishing House, 2011.

Zhu Ying. *Engineering Survey for Ballastless Track of Passenger Dedicated Railways.* Beijing: China Railway Publishing House, 2008.

Zhu Ying. Planning and Design of High-speed Railways. *High-speed Railway Technology*, 2010 (2): 1–5.

Zhu Ying. Devotion to Construction of High Speed Railway with Independent Intellectual Property Rights — Overall Design of Ballastless Track-integrated Pilot Section of SuiYu Railway. *Journal of Railway Engineering Society*, 2008 (supplement): 182–191.

Printed in the United States
By Bookmasters